Recent Advances in Sensor Array Signal Processing and Its Applications in Future Communication and Radar

Recent Advances in Sensor Array Signal Processing and Its Applications in Future Communication and Radar

Guest Editors

Xiaofei Zhang
Jianfeng Li
Wei Liu

MDPI

Basel • Beijing • Wuhan • Barcelona • Belgrade • Novi Sad • Cluj • Manchester

Guest Editors
Xiaofei Zhang
College of Electronic Information Engineering
Nanjing University of Aeronautics and Astronautics
Nanjing
China

Jianfeng Li
College of Electronic Information Engineering
Nanjing University of Aeronautics and Astronautics
Nanjing
China

Wei Liu
Department of Electrical and Electronic Engineering
The Hong Kong Polytechnic University
Hong Kong
China

Editorial Office
MDPI AG
Grosspeteranlage 5
4052 Basel, Switzerland

This is a reprint of the Special Issue, published open access by the journal *Sensors* (ISSN 1424-8220), freely accessible at: https://www.mdpi.com/journal/sensors/special_issues/ELW219343J.

For citation purposes, cite each article independently as indicated on the article page online and as indicated below:

Lastname, A.A.; Lastname, B.B. Article Title. *Journal Name* **Year**, *Volume Number*, Page Range.

ISBN 978-3-7258-5223-9 (Hbk)
ISBN 978-3-7258-5224-6 (PDF)
https://doi.org/10.3390/books978-3-7258-5224-6

Contents

About the Editors

Xiaofei Zhang

Xiaofei Zhang works at College of Electronic Information Engineering, Nanjing University of Aeronautics and Astronautics, and his research interests include array signal processing, mobile communication and source localization. He is the executive deputy director of the Key Laboratory of the Ministry of Industry and Information Technology, selected as a "Chinese Highly Cited Scholar" by Elsevier and one of the top 2% scientists in the world. He won the first prize of Natural Science from the China Electronics Society and the first prize of Technical Invention from the China Radar Industry Association.

Jianfeng Li

Jianfeng Li works at College of Electronic Information Engineering, Nanjing University of Aeronautics and Astronautics, and his research interests include array signal processing and source localization. He hosts projects like National Natural Science Foundation of China (Youth/General) and he won the first prize of Natural Science from the China Electronics Society, the first prize of Technical Invention from the China Radar Industry Association, and the first prize of National Defense Science and Technology Progress.

Wei Liu

Wei Liu received his BSc (Space Physics) and LLB (Intellectual Property Law) degrees from Peking University, China, in 1996 and 1997, respectively, in addition to his MPhil from the Department of Electrical and Electronic Engineering, University of Hong Kong in 2001, and PhD from the School of Electronics and Computer Science, University of Southampton, UK, in 2003. He then worked as a postdoc first at Southampton and later at Imperial College London. From 2005 to 2023, he was a Lecturer/Senior Lecturer at the Department of Electronic and Electrical Engineering, University of Sheffield, UK, and from 2023 to 2024 and a Reader at the School of Electronic Engineering and Computer Science, Queen Mary University of London (Visiting Professor for 2024-2027). Since 2024, he has been a Professor at the Department of Electrical and Electronic Engineering, Hong Kong Polytechnic University.

sensors

MDPI

Article

Outlier-Robust Three-Element Non-Uniform Linear Arrays Design Strategy for Direction of Arrival Estimation in MIMO Radar

Andrea Quirini *, Fabiola Colone and Pierfrancesco Lombardo

Department of Information Engineering, Electronics and Telecommunications DIET, Sapienza University of Rome, 00184 Rome, Italy; fabiola.colone@uniroma1.it (F.C.); pierfrancesco.lombardo@uniroma1.it (P.L.)
* Correspondence: andrea.quirini@uniroma1.it

Abstract: This paper presents a novel design strategy for outlier-robust, three-element non-uniform linear array (NULA) configurations optimized for multiple-input multiple-output (MIMO) radar systems aimed at target direction of arrival (DoA) estimation. The occurrence of outliers, i.e., ambiguous estimates, is a well-known issue in DoA estimation based on the maximum likelihood (ML), which is caused by the local maxima of the likelihood function. Specifically, we study how the positioning of both transmitters and receivers affects both presence of outliers and accuracy of ML DoA estimation. By leveraging a theoretical prediction of the DoA mean squared error (MSE), we propose a design strategy to jointly optimize the positions of NULA array of three transmitting and receiving elements, only inside a subspace which guarantees that the outlier probability remains below a specified threshold. Compared to NULA configurations with a single transmitter, the proposed designs achieve superior estimation accuracy due to two key factors: improved asymptotic performance resulting from a narrower mainlobe, and enhanced robustness against outliers due to reduced sidelobes. Furthermore, the proposed approach is well-suited for practical implementation in low-cost radars using only 3×3 or 2×3 MIMO configurations, as it also incorporates practical design constraints such as minimum inter-element spacing to account for the physical dimensions of the antennas, and tolerance in the installation accuracy.

Keywords: MIMO radar; AESA radar; non-uniform linear arrays; direction of arrival; maximum likelihood estimation

1. Introduction

Active electronically scanned array (AESA) radar architectures are well-established in radar surveillance, due to their enhanced flexibility compared to conventional radar systems. By using multiple transceiver modules, AESA radar systems allow for electronic control of the antenna beam direction, enabling rapid and precise scanning of the surveilled area without the need for physical movement of the antenna.

In recent years, the fully digital paradigm for AESA radar has gained attention in both industry and research, as it leverages digital signal processing to further enhance the flexibility of radar operations. Specifically, fully digital AESA radars aim to provide independent digital control of each antenna element for both transmission and reception. This is achieved by moving analog-to-digital converters (ADCs) and digital-to-analog converters (DACs) as close as possible to the antennas, allowing most signal processing operations to be implemented in the digital domain.

Next-generation fully digital arrays are currently employed in only a limited number of radar products, typically operating in single transmit mode. However, the independent control of transmit and receive antennas facilitates multiple-input multiple-output (MIMO) operations. Specifically, in fully digital AESA radar systems, each antenna can transmit different orthogonal waveforms directly synthesized in the digital domain, ensuring MIMO functionality and enabling flexible and efficient exploitation of the available antenna aperture.

A major challenge in MIMO radar systems is ensuring the orthogonality of waveforms, particularly when multiple transmit antennas are involved. Achieving orthogonality across a large number of waveforms over a sufficiently wide range and across the Doppler frequencies of interest is a nontrivial problem. Chirp signals, particularly linear frequency-modulated (LFM) signals, are known to offer quasi-orthogonal waveform pairs, such as the "chirp up" and "chirp down" pair. Additionally, further quasi-orthogonal waveforms can be designed by utilizing "V-chirp" signals with bidirectional sweeps.

In low-cost radar systems for surveillance, there is a strong interest in leveraging the potential of MIMO operation without significantly increasing system complexity. This requires the following:

(A) The use of a limited number of antennas to reduce the number of receiving chains and the volume of data to be processed.

(B) The use of a few simple orthogonal or quasi-orthogonal waveforms, such as chirps, which do not require complex optimization to ensure orthogonality.

In this perspective, a radar system with an extremely limited number of channels can be interesting for many low-cost surveillance applications, as it reduces costs and requires using a limited number of orthogonal waveforms. Systems with a few antennas are also beneficial for designing lightweight and compact systems, suitable for installation on moving platforms such as vehicles or aircraft, where the capability to process large amounts of data is also limited.

When aiming to use a limited number of antennas, non-uniform linear array (NULA) configurations are ideal. These configurations allow for the strategic positioning of the available sensors, including both short and long baselines with a small number of elements, therefore optimizing the trade-off between performance and cost. Specifically, this paper focuses on the application of target direction of arrival (DoA) maximum likelihood (ML) estimation, considering a MIMO radar system with three antennas. Three-element array configurations are widely used for the practical and cost-effective implementation of radar systems for DoA estimation (e.g., see [1–7]). This choice is driven by the target applications of our design strategy: low-cost, resource-constrained systems such as automotive radar or sensors for small unmanned aerial vehicles, where size, weight, power, and computational load are critical constraints. Each antenna can be either a transmitting/receiving element or a receiving-only element. Our goal is to optimize the positions of the three antennas to maximize the estimation accuracy, while constraining the probability of outliers among the obtained DoA estimates.

Achieving accurate DoA estimation with such a limited number of antenna elements is especially challenging. Specifically, using a typical uniform linear array (ULA) configuration with only three antennas would be highly constraining, leaving significant performance potential unexplored. In contrast, NULA configurations can be quite beneficial when there are severe constraints on the number of antenna elements, as they enable more flexible designs that can better exploit the limited resources available.

However, compared to ULA configurations, it is well known that NULA configurations introduce DoA estimation ambiguities in the direction of one of the sidelobes of the likelihood function. These ambiguities can significantly degrade DoA estimation accuracy, causing the DoA estimation mean square error (MSE) to deviate from the asymptotic performance, especially under limited signal-to-noise ratio (SNR) conditions. The accuracy of DoA estimation in the presence of outliers has been extensively studied in the open literature [8–17], and several lower bounds for the MSE have been derived, including the Barankin bound [10], the Bayesian Cramér–Rao Bound (CRB) [11–13], and the Ziv–Zakai bound [14,15].

The problem of DoA estimation has been extensively addressed in the literature, with powerful super-resolution algorithms like MUSIC [18], ESPRIT [19], and methods based on compressed sensing [20–22] as the classical choices. These super-resolution techniques allow resolving multiple closely-spaced targets. However, the work presented here focuses on a different challenge, predominant in low-cost, resource-constrained systems: ensuring robustness against estimation outliers for a single target at low SNR. For this purpose, the ML estimator is particularly suitable, as its performance is well-understood theoretically [16], allowing establishment of a direct relationship between array geometry and outlier probability, which is the core of our contribution.

Notably, Ref. [16] presented an accurate theoretical prediction for the DoA MSE in moderate to high SNR scenarios. Therein, it is shown that the probability of outliers can be expressed only in terms of the operative SNR and of the array geometry. This highlights the importance of the array geometry, which should be carefully designed to maximize the robustness against outliers, especially under limited SNR conditions. Several efforts have been undertaken for designing ambiguity-resistant NULA configurations over the years [23–29], including minimum redundancy arrays (MRA) [23] and ambiguity-resistant interferometers [24]. Particularly, in our previous work [29], we built upon the MSE theoretical prediction in [16] to devise a flexible design strategy for outlier-robust three-element NULA configurations on receive. The design strategy proposed therein was only applicable to SIMO radar systems, allowing us to control the probability of outliers for estimating the DoA of either emitting sources, targets illuminated by existing transmitters of opportunity (as in passive radar), or in scenarios where only one of the three antennas was equipped with a transmitter.

In this paper, we extend this outlier-robust design framework to co-located MIMO setups, which represents the main contribution of this work. This extension is non-trivial since the physical position of each antenna element simultaneously affects multiple virtual channels. Through theoretical derivation, we demonstrate that for the same physical placement of antennas, a MIMO configuration with multiple transmitters emitting orthogonal waveforms yields a different virtual array geometry, which generally results in an increased robustness against outliers. In turn, this modifies the array design strategy, enabling the design of longer-baseline NULAs that offer superior asymptotic accuracy for the same level of outlier robustness guaranteed by shorter configurations in the SIMO case.

Based on this analysis, we propose a NULA design strategy that allows system designers to fully exploit the limited number of available antennas, achieving an optimal balance between asymptotic accuracy and outlier robustness by leveraging the advantages of MIMO operation. In the paper, we assume independent digital control over each antenna element, as guaranteed by next-generation AESA systems, and orthogonality between waveforms—an assumption that can be approximately satisfied easily in practice with a system using a maximum of three transmitters, e.g., using quasi-orthogonal up, down, and V-shaped LFM waveforms.

This paper is organized as follows. Section 2 presents the MIMO signal model and revisits the MSE theoretical approximation derived in [16]. Section 3 presents the proposed NULA design strategy for MIMO radar, based on a constraint on the maximum outlier probability. Section 4 provides an example of an application of the proposed design strategy, considering practical technological constraints such as minimum array inter-element spacings or installation tolerance. Lastly, Section 5 draws our concluding remarks.

2. Array Signal Model and MSE Prediction

Consider the system geometry depicted in Figure 1, showing a NULA with N elements aligned along the x-axis. Let d_n denote the position of the n-th element. Without loss of generality, we assume that the leftmost element is at the origin of the axis, so that the array geometry is described by the vector $z = [d_0, \ d_1, \dots, d_{N-1}]$ collecting the positions of the array elements. Here, we assumed that the two-way phase center of the array element at the origin of the axis is the phase reference.

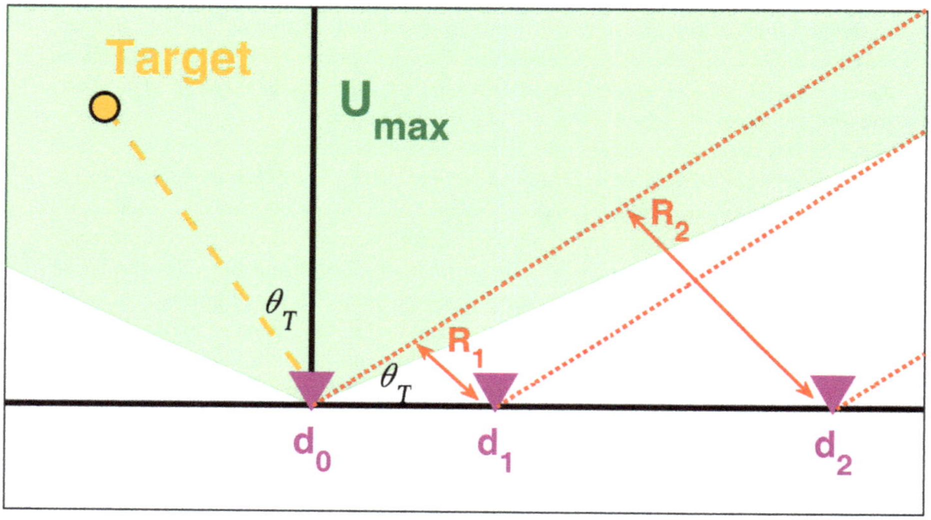

Figure 1. Array geometry and operative scenario.

We consider a target T located within the surveilled area at an angle θ_T relative to the array boresight. Our signal model focuses on an isolated target, assuming it has been resolved within a specific range resolution cell. The array comprises $M \leq N$ transmitting elements, each emitting a waveform denoted as $s_m[k]$. The signal contribution at the n-th antenna after matched filtering to the m-th waveform and at the target range bin can be expressed as:

$$x_{m,n} = A_{m,n} e^{-j2\pi \frac{d_m^{TX}}{\lambda} u_T} e^{-j2\pi \frac{d_n}{\lambda} u_T} + \sum_{i=0, i\neq m}^{M-1} A_{i,n} \rho_{m,\,i} e^{-j2\pi \frac{d_i^{TX}}{\lambda} u_T} e^{-j2\pi \frac{d_n}{\lambda} u_T} + w_{m,\,n}, \quad (1)$$

where

- $A_{m,n}$ represents the complex target signal amplitude after range compression;
- λ is the signal wavelength;
- $u_T \triangleq sin\theta_T$ is the target DoA;
- $\{d_m^{TX}, \ m = 0, \dots M-1\}$ is a subset of $\{d_n, \ n = 0, \cdots, N-1\}$;
- $\rho_{m,n}$ is the complex amplitude of the cross-correlation interference, which models the leakage from the i-th transmitted waveform into the m-th processed channel.
- $w_{m,n}$ is a complex-valued sequence representing the thermal noise contribution at the n-th receiving element;

However, for LFM waveforms with a high time-bandwidth product, such as those typically used in radar, the energy of the cross-talk term is significantly lower than the peak energy of the auto-correlation term. For this reason, its effect can be reasonably modeled as a small degradation of the effective SNR. Therefore, the theoretical derivations are obtained assuming perfect orthogonality, but the impact of this non-ideal behavior will be considered in simulated examples later in the paper.

To obtain an estimate $\hat{u}_T$ of the target DoA u_T, we resort to the ML estimation technique. A block diagram of the fundamental processing scheme considered in this paper is shown in Figure 2. Specifically, the received signals at different antennas are first range-compressed via matched filtering, which involves multiplying the received signal spectrum $X_n[p]$ by the complex conjugate of the spectrum of each transmitted signal, $S_m[p]$. This is performed by the block named "MF and TX separation", which is exploded in its details only for channel k = 1 for compactness. This operation enables both target range estimation and separation of the contributions from each transmitter at each receiver. For simplicity, in Figure 2, the range compression process is represented assuming that $M = 3$ transmitters are present. In cases with $M = 1$ or $M = 2$ transmitters, the processing scheme is simplified accordingly.

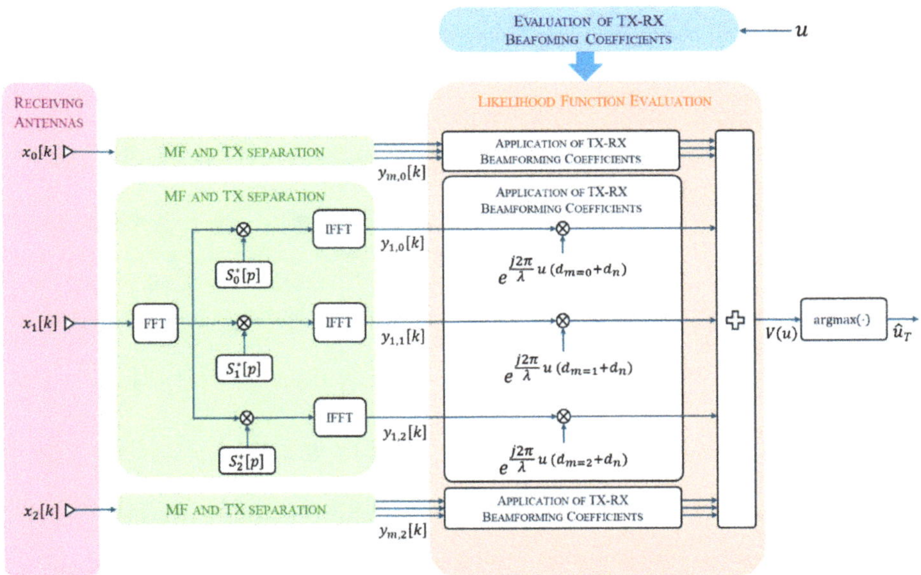

Figure 2. Block diagram of the MIMO processing scheme.

To carry out a fair comparison of SIMO and MIMO, we assume that the total transmitted power is the same for every setup. Therefore, with M transmitters, the amplitudes $A_{m,n}$ are scaled by $\sqrt{M}$, to account for the power distribution across the array.

Following the processing scheme in Figure 2, each range-compressed sequence $y_{m,n}[k]$ undergoes a beamforming stage, where the transmit and receive phase shifts are compensated to estimate the target DoA. To this end, Equation (1) can be conveniently rewritten in vector notation as follows:

$$\boldsymbol{x} = \boldsymbol{A} \odot \boldsymbol{s}(u_T) + \boldsymbol{w}, \tag{2}$$

where

- $\boldsymbol{x} = [x_{1,1}, \ldots, x_{1,N}, x_{2,1}, \ldots x_{2,N}, \ldots x_{M,1} \ldots, x_{M,N}]^T$ is a column vector of size $(NM \times 1)$, collecting the signal contributions $x_{m,n}$;

- $\boldsymbol{A} = [A_{1,1}, \ldots, A_{1,N}, A_{2,1}, \ldots A_{2,N}, \ldots A_{M,1} \ldots, A_{M,N}]^T$ is a column vector of size $(NM \times 1)$, collecting the complex amplitudes $A_{m,n}$;
- $\boldsymbol{s}(\theta_T)$ is a steering vector of size $(NM \times 1)$, collecting the phase shifts $e^{-j2\pi \frac{d_m^{TX}+d_n}{\lambda} u_T}$ due to the two-way path;
- $\boldsymbol{w} = CN(0, \sigma_n^2)$ is a column vector of size $(NM \times 1)$, collecting the MN thermal noise samples, drawn from a complex Gaussian distribution with zero mean and σ_n^2;
- $\odot$ denotes the elementwise Hadamard product.

Based on this signal model, we can define an ideal likelihood function for our ML estimator as in [11]:

$$V(u) = \left| \boldsymbol{s}^H(u)\boldsymbol{x} \right|^2, \tag{3}$$

Here, $\boldsymbol{s}^H(u)$ represents the steering vector corresponding to the generic search direction $u = sin\theta$. Assuming the target is within a specific angular region of interest, denoted as $U_{MAX} = [-u_{max}, u_{max}]$, the ML estimator is given by:

$$\hat{u}_T = \max_{u \in U_{MAX}} \{V(u)\}, \tag{4}$$

A theoretical performance prediction for the MSE of the ML estimator has been derived in [16], and it is briefly recalled here for the reader convenience. This performance prediction is based on the hypothesis that the MSE contribution can be decomposed in the sum two main terms:

- A small error contribution, due to distortions on the likelihood function main lobe caused by thermal noise;
- A large error contribution due to the presence of the outliers, namely DoA estimates located well outside the main lobe peak on the likelihood function, around the angles, u_m, $m = 1, \ldots, M$ where it has its sidelobes'. These errors are still caused by the presence of thermal noise that occasionally makes them appear higher than the mainlobe peak.

Specifically, the MSE prediction $\zeta(SNR, d_1, d_2, u_T)$ is decomposed in [16] as:

$$\zeta(SNR, d_1, d_2, u_T) \triangleq E\left[(\hat{u}_T - u_T)^2\right] \approx \left[1 - \sum_{m=1}^{M} P_m\right] \cdot CRB + \sum_{m=1}^{M} (u_m - u_T)^2 P_m, \tag{5}$$

where

- M represents the number of sidelobe peaks in the angular sector of interest U_{MAX};
- $P_m = P_m(SNR, d_1, d_2, u_T)$ represents the mth pairwise error probability, i.e., the probability that the mth sidelobe peak is higher than the main lobe peak;
- u_m is the m-th sidelobe peak location;
- CRB is the Cramer–Rao lower bound, representing the MSE achieved when the contribution due to outliers is negligible.

For this approximation to be tight, the union bound limit must apply [25,30], meaning that the probability of two side peaks simultaneously being higher than the main lobe peak must be negligible. Clearly, this depends on both the input SNR and the likelihood function's side peaks amplitudes, which in turn depend on the array geometry. In practice, for a given array geometry, there is a specific SNR value above which the MSE approximation is tight.

This SNR value also determines the lower limit of the so-called threshold region [16], where the probability of outliers is non-negligible, and the achieved MSE deviates from the CRB. When the input SNR is high enough, the probability of outliers becomes negligible, making the CRB a good approximation of the MSE. Since in this case the asymptotic performance is reached, this SNR region is referred to as the asymptotic region in [16]. Lastly, when the SNR is too low, the union bound approximation is no longer valid, meaning that there is a non-negligible probability of multiple side peaks being higher than the main lobe peak. Under this condition, the DoA can no longer be estimated, so this SNR region is referred to as the no-information region in [16]. A qualitative overview of the three regions described is shown in Figure 3.

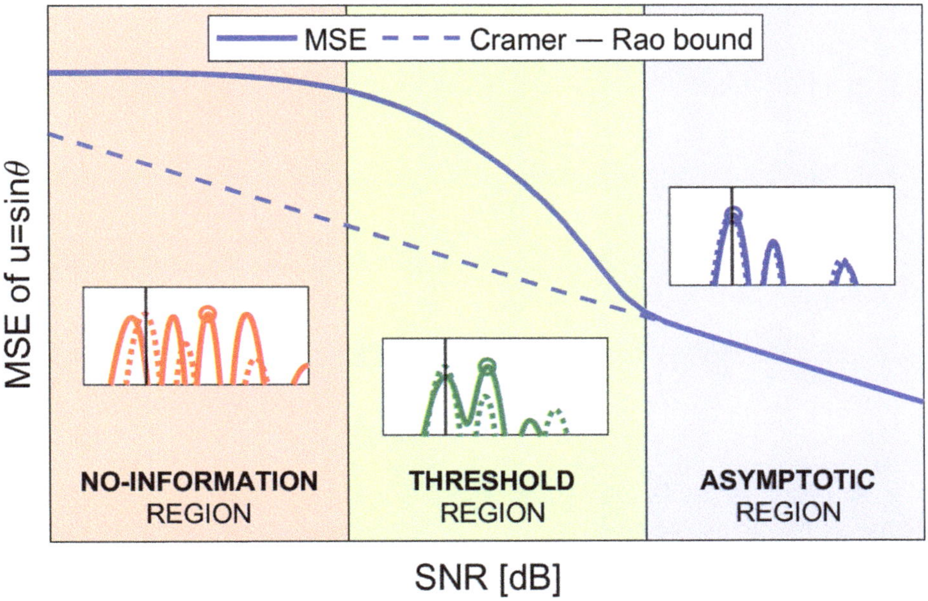

Figure 3. Qualitative behavior of the mean square error (MSE) vs. signal-to-noise power ratio (SNR).

The goal of this paper is to control the probability of outliers in the estimates obtained by the MIMO radar by optimizing the design of the three-element array. Specifically, we aim to jointly optimize the positions of both the transmitting/receiving elements and the receiving-only elements, enabling the design of MIMO NULA configurations, which are robust to outliers, not only in the asymptotic region, but also for SNR values within the threshold region. To this end, in the next section we constrain the maximum value for the MIMO pairwise error probabilities P_m in Equation (5), aiming to reduce the contribution of outlier estimates to the global MSE.

3. Admissible Pairwise Probability Region

In this section, we carry out the analytical derivations required to design outlier-robust NULA configurations, tailored for MIMO radar systems. From Equation (5), we note that the m-th pairwise error probability P_m acts as a scaling factor for the large error contributions $(u_m - u_T)^2$ that cause the MSE to deviate from the CRB. Therefore, in the following we will call outlier-robust array configurations all those NULAs which satisfy a desired constraint on the maximum pairwise error probabilities P_m. Based on [16], the mathematical expression for the m-th pairwise error probability is:

$$P_m = Q(a,b) - \frac{1}{2}e^{-\frac{SNR}{2}} I_0\left(\frac{SNR}{2}\frac{\sqrt{L(u)}}{MN}\right), \tag{6}$$

where

- I_0 denotes the modified Bessel function of the first kind and order zero;
- SNR is the signal-to-noise power ratio, directly related to the complex baseband a target amplitude, i.e., $SNR = A^2$;
- $Q(a, b)$ denotes the Marcum Q function of the two arguments

$$a \triangleq \sqrt{\frac{SNR}{2} M\left(1 - \sqrt{1 - \frac{L(u)}{M^2N^2}}\right)},$$
$$b \triangleq \sqrt{\frac{SNR}{2} M\left(1 + \sqrt{1 - \frac{L(u)}{M^2N^2}}\right)},$$ (7)

and

$$L(u) = \left|s^H(u)s(u_T)\right|^2$$ (8)

is the normalized noiseless likelihood function that depends only on the array geometry and achieves the maximum value of MN when $u = u_T$.

The first step of the design strategy is to impose a constraint on the maximum value of P_m, namely:

$$P_m \leq P_{max}, \forall m = 1, \ldots, M.$$ (9)

From Equation (6), we note that P_m only depends on the array geometry, by means of the noiseless likelihood function, and has a monotonic behavior with the operational SNR, which determines how much this noiseless likelihood function gets distorted. Therefore, for a fixed SNR value, the constraint on the maximum admissible P_m value can be converted into a constraint on the maximum admissible value for the m-th sidelobe peak of the normalized likelihood, namely $L(u_m) \leq L_{max}$, where the maximum admissible value L_{max} can be univocally determined by inverting Equation (6) for a given SNR value.

The problem then becomes to identify all the array configurations with M transmit elements and N receive elements that satisfy the constraint $L(u_m) \leq L_{max}$. To this end, we define the angles

$$\phi_k \triangleq 2\pi \frac{d_k}{\lambda}(u - u_T),$$ (10)

and express $L(u)$ for the generic N and M as:

$$L(u) = \left|\sum_{m=0}^{M-1}\sum_{n=0}^{N-1} e^{j2\pi\frac{d_m^{TX}+d_n}{\lambda}(u-u_T)}\right|^2 = \Psi_M\left(\phi_0^{TX}, \cdots, \phi_{M-1}^{TX}\right) \cdot \Psi_N(\phi_0, \cdots, \phi_{N-1}),$$ (11)

where for the generic number of elements K:

$$\Psi_K(\phi_0, \cdots, \phi_{K-1}) = \left|\sum_{k=0}^{K-1} e^{j\phi_k}\right|^2$$ (12)

In the following, taking the phase reference in the most left-side element $n = 0$, it is assumed for simplicity $\phi_0 = 0$, so that ϕ_0 is not listed among the variables of the functions Ψ_N and Ψ_M.

As is apparent, we successfully decoupled the contributions of the phase shifts due to the transmitter-to-target path, encapsulated in $\Psi_M(\phi_0^{TX}, \cdots, \phi_{M-1}^{TX})$, from those due to the target-to-receiver path, encapsulated in $\Psi_N(\phi_0, \cdots, \phi_{N-1})$. Based on Equation (11), we can derive the outlier-robust NULA design strategy for MIMO radar configurations. In the following subsections, three different setups are considered:

- A 3×3 MIMO setup with three trans-receiving array elements;

- A 2 × 3 MIMO setup with two trans-receiving array elements, and a third receiving-only element;
- A SIMO setup with a single trans-receiving element and two receiving-only elements.

3.1. 3 × 3 MIMO with Three Trans-Receiving Array Elements

Based on Equation (11), in the 3 × 3 MIMO setup, $N = M = 3$, $d_m^{TX} = d_m$, $m = 0, \cdots, M-1$, and the transmission functional is identical to the reception functional, so that the constraint $L(u) \leq L_{max}$ becomes:

$$L(\phi_1, \phi_2) = \left|1 + e^{j\phi_1} + e^{j\phi_2}\right|^4 \tag{13}$$

We notice that this functional has a large number of symmetries among its angles, among which:

$$L(\phi_1,\phi_2) = L(\phi_2,\phi_1) = L(\phi_2-\phi_1,\phi_2) = L(\phi_1-\phi_2,-\phi_2) = L(-\phi_1,\phi_2-\phi_1) = L(\phi_1,\phi_1-\phi_2) \tag{14}$$

By defining a 45° rotated axis system as:

$$\begin{cases} x = \frac{\phi_2-\phi_1}{2} \\ y = \frac{\phi_1+\phi_2}{2} \end{cases}, \begin{cases} \phi_1 = y - x \\ \phi_2 = y + x \end{cases} \tag{15}$$

and following the derivations in Appendix A, we obtain the following mathematical representation of the set of points identified by cutting the normalized likelihood function at the value L_{max} in the (ϕ_1, ϕ_2) plane.

$$\begin{cases} \phi_1 = \pm acos\left[\frac{\sqrt{L_{max}}-1}{4cosx} - cosx\right] + 2\pi h - x \\ \phi_2 = \pm acos\left[\frac{\sqrt{L_{max}}-1}{4cosx} - cosx\right] + 2\pi k + x \end{cases}, \tag{16}$$

with the constraint

$$-acos\left(\frac{-1+\sqrt[4]{L_{max}}}{2}\right) < x \leq acos\left(\frac{-1+\sqrt[4]{L_{max}}}{2}\right). \tag{17}$$

A contour plot of the likelihood function for the 3 × 3 MIMO case is shown in Figure 4d.

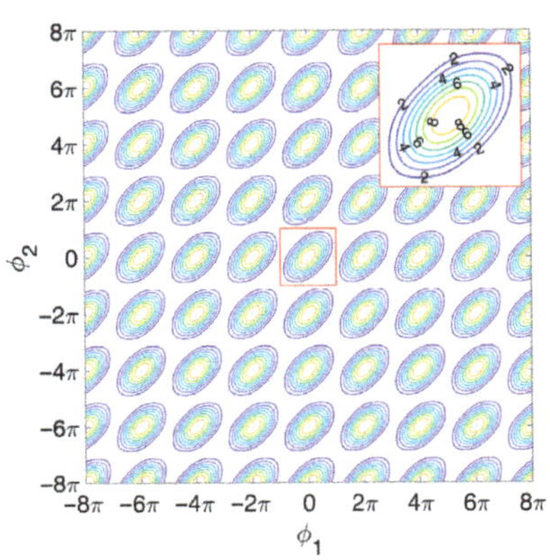

(**a**) SIMO

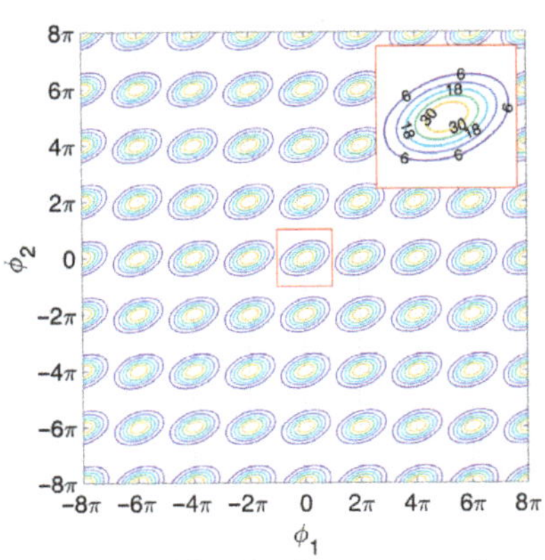

(**b**) 2 × 3 MIMO with adjacent transmitters

Figure 4. *Cont.*

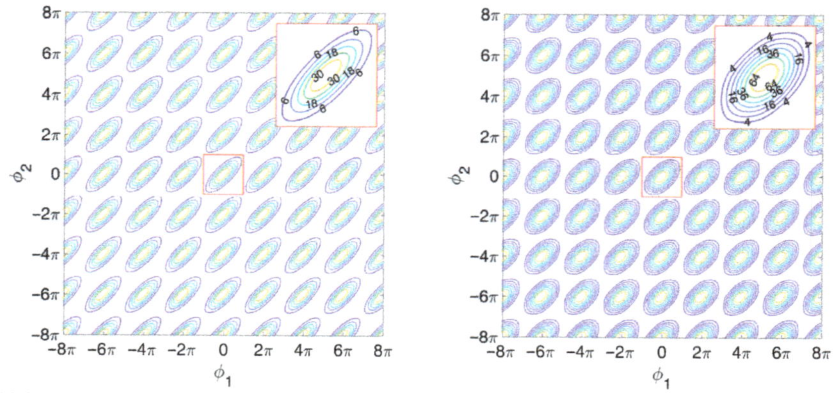

(**c**) 2 × 3 MIMO with transmitters at the edges (**d**) 3 × 3 MIMO

Figure 4. Contour plot of the noiseless normalized likelihood function $L(\phi_1, \phi_2)$: (**a**) SIMO; (**b**) 2 × 3 MIMO with adjacent transmitters; (**c**) 2 × 3 MIMO with external transmitters; (**d**) 3 × 3 MIMO.

3.2. 2 × 3 MIMO with Two Trans-Receiving Array Elements

The 2 × 3 MIMO case can be divided into three subcases, where the single receiving-only element is positioned alternately at the leftmost, rightmost, or central position. To mathematically represent the constraint $L(u) \leq L_{max}$ in the (ϕ_1, ϕ_2) plane when only two of the array elements are equipped with a transmitter, we conveniently set the phase reference at the receiving-only element, as this simplifies the derivation. The three subcases of the 2 × 3 MIMO configuration can be readily obtained from one another through a simple coordinate transformation in the (ϕ_1, ϕ_2) plane.

By setting $M = 2$, $d_0^{TX} = d_1$, $d_1^{TX} = d_2$, and taking the phase reference at the receiving-only element the transmission functional for the 2 × 3 MIMO case is given by:

$$\Psi_2\left(\phi_0^{TX}, \phi_1^{TX}\right) = \Psi_2(\phi_1, \phi_2) = \left|e^{j\phi_1} + e^{j\phi_2}\right|^2, \tag{18}$$

while the reception functional is the same as in the 3 × 3 MIMO case. Therefore, we obtain:

$$L(u) = \Psi_2\left(\phi_0^{TX}, \phi_1^{TX}\right) \cdot \Psi_3(\phi_1, \phi_2) = \left|e^{j\phi_1} + e^{j\phi_2}\right|^2 \left|1 + e^{j\phi_1} + e^{j\phi_2}\right|^2 \leq L_{max}. \tag{19}$$

The derivation details are provided in Appendix B, while here we only report the expressions to mathematically characterize the set of points in (ϕ_1, ϕ_2), identified by cutting the normalized likelihood function at L_{max}:

$$\begin{cases} \phi_1 = \pm acos\left[\dfrac{\frac{L_{max}}{4cos^2 x} - 1}{4cosx} - cosx\right] + 2\pi h - x \\ \phi_2 = \pm acos\left[\dfrac{\frac{L_{max}}{4cos^2 x} - 1}{4cosx} - cosx\right] + 2\pi k + x \end{cases}, \tag{20}$$

with

$$-acos\left(-\frac{1 - \sqrt{1 + 4\sqrt{L_{max}}}}{4}\right) < x \leq acos\left(-\frac{1 - \sqrt{1 + 4\sqrt{L_{max}}}}{4}\right), \tag{21}$$

Finally, we describe how to transition between different 2 × 3 MIMO configurations.

For the 2 × 3 MIMO case with the receiving-only antenna at the rightmost array element, using $M = 2$, $d_0^{TX} = d_0$, $d_1^{TX} = d_1$ in Equation (12) yields:

$$\Psi_2\left(\phi_0^{TX}, \phi_1^{TX}\right) = \Psi_2(\phi_1) = \left|1 + e^{j\phi_1}\right|^2, \tag{22}$$

while the reception functional is the same as in the square root of the 3 × 3 MIMO case. Therefore, we obtain:

$$L(u) = \Psi_2\left(\phi_0^{TX}, \phi_1^{TX}\right) \cdot \Psi_3(\phi_1, \phi_2) = \left|1 + e^{j\phi_1}\right|^2 \left|1 + e^{j\phi_1} + e^{j\phi_2}\right|^2 \leq L_{max}. \quad (23)$$

It is easy to notice that the configuration is perfectly symmetric to the previous one. Therefore, the resulting coordinate transformation is straightforward. Specifically, by defining $\phi_1' \triangleq \phi_2 - \phi_1$, $\phi_2' \triangleq \phi_2$, we can transform the TX functional in the TX functional of the previous case, while the RX functional remains unchanged thanks to its symmetry properties listed above.

Based on this coordinate transformation, we can reuse Equations (20) and (21) to describe the set of points in the (ϕ_1, ϕ_2) plane that correspond to the normalized likelihood function being cut at L_{max} in the 2 × 3 MIMO case with a receiving-only antenna at the rightmost element.

Similarly, for the 2 × 3 MIMO case where the external elements of the array are used as transmitters and the receiving-only element is at the center, using $M = 2$, $d_0^{TX} = d_0$, $d_1^{TX} = d_2$ in Equation (12) yields:

$$\Psi_2\left(\phi_0^{TX}, \phi_1^{TX}\right) = \Psi_2(\phi_1) = \left|1 + e^{j\phi_1}\right|^2, \quad (24)$$

while the reception functional is the same as in the square root of the 3 × 3 MIMO case. Therefore, we obtain:

$$L(u) = \Psi_2\left(\phi_0^{TX}, \phi_1^{TX}\right) \cdot \Psi_3(\phi_1, \phi_2) = \left|1 + e^{j\phi_2}\right|^2 \left|1 + e^{j\phi_1} + e^{j\phi_2}\right|^2 \leq L_{max}. \quad (25)$$

Although this case has no direct relationship with the previous ones, we notice that it can be obtained from the first one by defining $\phi_1'' \triangleq \phi_2 - \phi_1$, $\phi_2'' \triangleq -\phi_1$, which again transforms Equation (24) in the TX functional of the first case, while the RX functional remains unchanged thanks to its symmetry properties listed above. Contour plots of the likelihood function in the 2 × 3 MIMO cases are shown in Figure 4b,c. Specifically, in Figure 4b, we assume adjacent transmitters, with the receiving-only antenna located at the leftmost array element. The symmetric case is not included, as the likelihood function appears identical. Furthermore, Figure 4d shows the likelihood function contour plot obtained in the 2 × 3 MIMO case with transmitters located at the edges of the array.

3.3. SIMO with a Single Trans-Receiving Array Elements

Finally, we briefly describe how the mathematical formulation of the constraint $L(u) < L_{max}$ changes in the SIMO case. A detailed derivation along these lines is provided in our previous work [29]. Specifically, when only a single transmitter is present ($M = 1$), and if the transmitter corresponds to the leftmost array element, M = 1, $d_0^{TX} = d_0$, the transmission functional $\Psi_M(\phi_m)$ simplifies to:

$$\Psi_1(\phi_0) = \left|e^{j\phi_0}\right|^2 = \Psi_1 = 1. \quad (26)$$

Therefore, in the SIMO case the normalized likelihood function only depends on the reception functional $\Psi_N(\phi_n)$:

$$L(u) = \Psi_N(\phi_n) = \left|1 + e^{j\phi_1} + e^{j\phi_2}\right|^2 \leq L_{max}. \quad (27)$$

Here, the normalized likelihood function is equal to the square root of the one in Equation (13), obtained in the 3 × 3 MIMO case. Therefore, the mathematical representation of the set of points identified by cutting the normalized likelihood function at L_{max} is obtained from that of the SIMO case by substituting $\sqrt{L_{max}}$ with L_{max}:

$$\begin{cases} \phi_1 = \pm acos\left[\frac{L_{max}-1}{4cosx} - cosx\right] + 2\pi h - x \\ \phi_2 = \pm acos\left[\frac{L_{max}-1}{4cosx} - cosx\right] + 2\pi k + x \end{cases}, \quad (28)$$

with

$$-acos\left(\frac{-1+\sqrt{L_{max}}}{2}\right) < x \leq acos\left(\frac{-1+\sqrt{L_{max}}}{2}\right). \quad (29)$$

Figure 4d shows the likelihood function contour plot obtained in the SIMO case.

By comparing the contour plots in Figure 4, we note that the different likelihood functions obtained for the considered MIMO cases appear distorted compared to the one for SIMO. This is since the dependency of the $\Psi_M(\phi_1, \phi_2)$ functional on ϕ_1 and ϕ_2 changes depending on the transmitter positioning. Notably, the 3 × 3 MIMO (Figure 4d) has the same shape as the one for the SIMO (Figure 4a), since in this case the transmission and reception functionals are the equal, so that the 3 × 3 MIMO likelihood function is simply the square of the SIMO likelihood function.

3.4. From the Likelihood Function Contours to the Admissible Region

After describing the likelihood functions for all the MIMO design case studies, we need to convert the derived mathematical representations into suitable admissible regions in the inter-element distance plane. This can be done by following a simple procedure, similar to the one in [29]. Particularly, Equation (10) allows us to map the quantities ϕ_1 and ϕ_2 into inter-element distances. This requires knowing the maximum range of variation of $(u - u_T)$, which depends on the extremes of the surveilled angular sector U_{max}. Specifically, we have:

$$-u_{max} \leq u - u_T \leq u_{max}. \quad (30)$$

The value of u_{max} acts as a scale factor for the axes, allowing us to rewrite ϕ_1 and ϕ_2 in terms of the two inter-element distances:

$$\frac{d_k}{\lambda} = \frac{\phi_k}{2\pi} \cdot \frac{1}{u_{max}}. \quad (31)$$

After this scaling, the mathematical representation of the set of points identified by the cut of the normalized likelihood function at the value L_{max} can be used to obtain an admissible pairwise probability (APP) region in the plane of inter-element distances including all the array configurations satisfying the constraint $L(u_m) < L_{max}$.

The example in Figure 5 allows us to better clarify the procedure to obtain the APP region.

We note that the locus of points corresponding to the array beampattern of a specific 3-element array $\bar{z} = \left[0, \bar{d}_1, \alpha\bar{d}_1\right]$ can be represented on the (ϕ_1, ϕ_2) plane as a linear segment $\overline{NP}$, as shown in Figure 5. The length of the segment is related to the total length of the array $\bar{z}$, while its slope α is determined based on the relationship between the two inter-element distances $\bar{d}_1$ and $\bar{d}_2$. Specifically, based on Equation (10), we have:

$$\alpha \triangleq \frac{\bar{\phi}_2}{\bar{\phi}_1} = \frac{\bar{d}_2}{\bar{d}_1}, \quad (32)$$

so that, for instance, all ULA configurations correspond to segments that lie on the line with slope $\alpha = 2$, since for every three-element ULA we have $d_2 = 2d_1$.

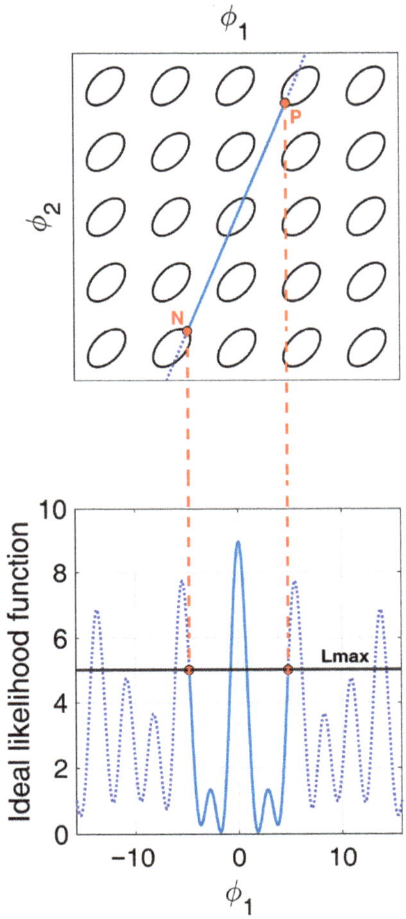

Figure 5. Likelihood function $L(\phi_1, \phi_2)$ cut at L_{max}. The linear segment $\overline{NP}$ with slope α represents the beampattern of a specific 3-element array $\bar{z} = \left[0, \bar{d}_1, \alpha\bar{d}_1\right]$.

For a fixed value L_{max}, for increasing values of ϕ_1, the first intersection with $L(\phi_1, \phi_2)$ along the slope α represents a specific solution $\left(\overline{\phi}_1, \overline{\phi}_2\right)$ of the equation

$$L(\phi_1, \phi_2) = L_{max}. \tag{33}$$

For values of ϕ_1 greater than $\overline{\phi}_1$ along the slope α, the constraint $L(\phi_1, \phi_2) \leq L_{max}$ is violated. This is visible in Figure 5, where the beampattern of the array $\bar{z}$ is shown as a function of ϕ_1.

Based on the considerations made in Figure 5, the APP region can be obtained by following the procedure outlined in [29], and recalled here:

I. Inside the set of all possible slopes A $= [1, +\infty]$, determine the subset of slopes α blocked by the first-order replica of $L(\phi_1, \phi_2)$, i.e., $(k, h) = (0, 1)$. Notice that all the replicas $(k, h) = (0, h)$ are blocked by replica (0,1), so they will be ignored in the following.

II. Update the set of the non-blocked slopes A after step I.

III. Inside the set A, determine the subset of slopes α blocked by the next higher-order replica of $L(\phi_1, \phi_2)$, i.e., $(k,h) = (1,1)$. Notice that the replicas $(k,h) = (h,h)$ are blocked by replica (1,1), so they will be ignored in the following.
IV. Update the set of the non-blocked slopes A after step III.
V. Inside the set A, determine the subset of slopes α blocked by higher-order replicas of $L(\phi_1, \phi_2)$.
VI. Update the set of the non-blocked slopes A after step V.
VII. Repeat steps V and VI until the set A is empty.

The presented procedure could be extended to $N = 4$ antenna elements, by operating in a 3D space, where the contours are replaced by surfaces. While the extension is feasible, the analytical derivation would make the description cumbersome. Therefore, we avoid an explicit illustration of this case and leave it to the interested reader.

Using this procedure, the likelihood function contours in Figure 4 can be easily converted into equivalent APP regions, containing all the array configurations that satisfy the constraint $P_m \leq P_{max}$. Figure 6a–d show the admissible regions obtained from the likelihoods in Figure 4, in the 2D domain $(d_1, d_2 - d_1)$ of the two inter-element distances, assuming $SNR = 15\ dB$, $P_{max} = 1 \cdot 10^{-3}$, and $U_{max} = [-60°, 60°]$, along with the likelihood function cut at L_{max}. As expected, the APP region appears symmetrical in this domain when using the 3×3 MIMO configuration, the 3×1 SIMO configuration and the 3×2 MIMO with the external transmitting elements. This is clearly not the case for the 3×2 MIMO configurations with adjacent transmitting elements, because the transmitters insisting only on one of the two baselines makes the behavior unsymmetric.

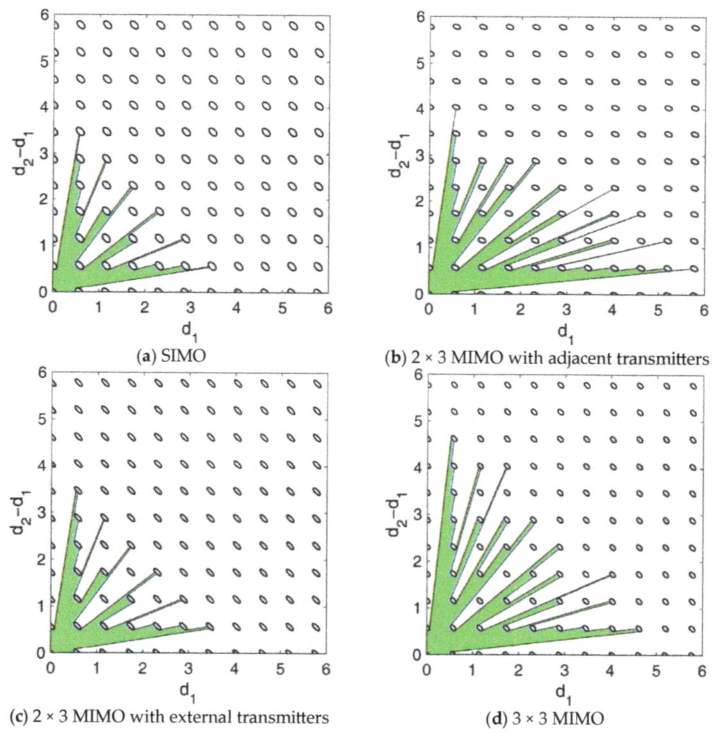

Figure 6. Admissible regions obtained assuming $SNR = 15\ dB$, $P_{max} = 1 \cdot 10^{-3}$, $U_{max} = [-60°, 60°]$. (**a**) SIMO; (**b**) 2×3 MIMO with adjacent transmitting elements (the transpose applies when using elements 1 and 2 as TX instead of elements 0 and 1); (**c**) 2×3 MIMO with external transmitting elements; (**d**) 3×3 MIMO.

In conclusion, in this section, we derived the mathematical expression to identify an APP region in the plane of the two inter-element distances for each MIMO case considered. The derived APP region contains all the NULA configurations that satisfy the constraint imposed on the pairwise error probabilities, allowing us to design outlier-robust three-element NULA configurations based on the design parameters, namely the operational SNR, the maximum pairwise error probability, and the angular sector of interest. An example of application of the proposed strategy is shown in the next section, where we compare the DoA estimation accuracy and the robustness to outliers of different NULA configurations resulting from the proposed design strategy.

4. A Design Case Study of the Proposed Strategy

In this section, we show an application example of the design strategy introduced in the previous section, and we compare different NULA designs considering both SIMO and MIMO operation. Particularly, we assume the same design parameters used in the previous section, namely $SNR = 15$ dB, $P_{max} = 1 \cdot 10^{-3}$, and $U_{max} = [-60°, 60°]$. We further assume transmitting narrowband orthogonal signals in the Wi-Fi bandwidth, with a carrier frequency of $f_0 = 2.4$ GHz, corresponding to a wavelength of $\lambda \approx 0.125$ m.

Before proceeding with the design, we note that the results in this section also relax the assumption of perfect waveform orthogonality. Specifically, while the theoretical MSE curves are obtained using the theoretical prediction derived in [16] and reported in Equation (5), they are compared with Monte Carlo simulation results. For these simulations, we considered a set of quasi-orthogonal LFM waveforms (up-chirp, down-chirp, and V-chirp), each with a time-bandwidth product of 200 ($B = 20$ MHz, $T = 10\mu$s). This allows us to demonstrate that the non-perfect orthogonality between waveforms does not significantly alter the validity of our theoretical analysis. Specifically, due to the high compression gain, the energy of the residual cross-talk after range compression is negligible compared to the main auto-correlation peak.

Similar to [29], the proposed design strategy is particularly flexible, as the identified APP region eventually allows the designer to account for additional technological constraints that may emerge during the design phase. Particularly, in this section, the APP regions resulting from the design procedure are further adjusted to account for two additional technological constraints:

- Minimum inter-element distance: To account for the physical size of the antennas (assumed to be 14 cm), we enforce a minimum inter-element distance of $l \approx 1.12\lambda$. This constraint, which excludes the blue area in Figure 7 from the APP region, carries the important secondary benefit of mitigating mutual coupling, since the electromagnetic interaction between elements is significantly reduced at large element separations.
- Mounting tolerance: Since the antennas cannot be positioned with infinite precision, we assume a mounting tolerance of 1.5 mm in each direction, corresponding to $\delta = 0.012\lambda$ in wavelength units. This requires excluding from the APP region the areas close to the region border (i.e., the red region in Figure 7).

These two constraints modify the admissible regions in Figure 6 as shown in Figure 7. The following observations are in order:

- As shown in Figure 6a, the SIMO configuration already achieved the smallest APP region compared to MIMO setups (in Figure 6b–d). Therefore, the additional technological constraints in this case severely limit the set of admissible NULA configurations, as the residual APP region (Figure 7a) becomes extremely small, leaving very little to no flexibility in the array design choice.
- By bringing the available two-way phase centers closer together, the MIMO 2×3 with adjacent transmitters prioritizes outlier robustness. This results in a larger APP region

even after technological constraints (Figure 7b), providing significant flexibility and greater design freedom compared to SIMO.

- In contrast, the MIMO 2 × 3 with transmitters at the edges of the array (Figure 7c) increases the overall separation between the array two-way phase centers, resulting in a narrower main lobe in the likelihood function. This leads to improved performance in the high SNR conditions, prioritizing asymptotic estimation accuracy over outlier robustness. Therefore, the resulting APP region is very similar to the one obtained in the SIMO case.
- Finally, the MIMO 3 × 3 (Figure 7d) benefits from the use of three transmitters, providing more balanced performance. Since the available power now is equally split among nine different two-way phase centers, we can benefit from both improved asymptotic estimation accuracy and enhanced robustness to outliers, resulting in the largest admissible region.

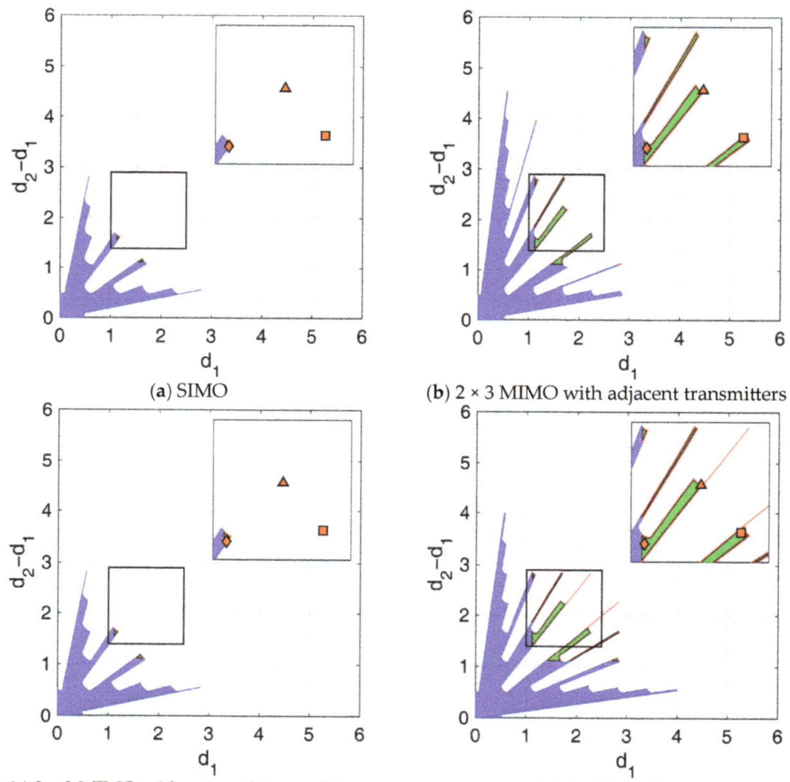

(**a**) SIMO (**b**) 2 × 3 MIMO with adjacent transmitters

(**c**) 2 × 3 MIMO with external transmitters (**d**) 3 × 3 MIMO

Figure 7. Admissible regions obtained assuming $SNR = 15$ dB, $P_{max} = 1 \cdot 10^{-3}$, $U_{max} = [-60°, 60°]$, and considering additional technological constraints (antenna size: 14 cm, mounting tolerance: 1 cm). (**a**) SIMO; (**b**) 2 × 3 MIMO with adjacent transmitters; (**c**) 2 × 3 MIMO with transmitters at the edges; (**d**) 3 × 3 MIMO.

Based on the above observations, we conduct two different types of comparisons between the different SIMO and MIMO configurations:

1. Comparison with a fixed array configuration: We analyze the performance of the NULA $\mathbf{z}_1 = [0\ 1.16\ 2.77]\lambda$ ('◊' marker), which is admissible in all SIMO/MIMO APP regions.

2. Comparison with a fixed maximum pairwise error probability: We compare the performance of $z_1 = [0\ 1.16\ 2.77]\lambda$ ('◊' marker) for the SIMO and the 2×3 MIMO configuration with transmitters at the edges, $z_2 = [0\ 2.20\ 3.91]\lambda$ ('□' marker) for the 2×3 MIMO with adjacent transmitters, and $z_3 = [0\ 1.76\ 3.98]\lambda$ ('Δ' marker) for the 3×3 MIMO. All configurations ensure a maximum pairwise error probability of approximately $P_{max} \approx 0.6 \cdot 10^{-3}$.

The first comparison highlights two key advantages of MIMO: (i) the ability to control outlier probability and (ii) the improved asymptotic performance of MIMO configurations compared to SIMO, which results from an increase in the total effective array length. Specifically, Figure 8a shows the MSE of the different setups, averaged within the U_{max} sector, as a function of SNR, while Figure 8c shows the probability of outlier and the maximum pairwise error probability as a function of the SNR. The following observations are in order:

- When the SNR is within the threshold region, the MSE obtained with array z_1 progressively improves from SIMO to MIMO. Specifically, the 3×3 MIMO achieves the lowest MSE. Instead, the MSE of the 2×3 MIMO configuration with external transmitters is relatively close to the SIMO, since the probability of an outlier is still significant at this SNR value. This is also clearly visible from the red curve Figure 8c.
- As the SNR increases and the performance gets closer to the asymptotic region, the impact of outliers on the MSE becomes negligible (see Figure 8d). As a result, the asymptotic term in Equation (6) becomes the dominating MSE contribution. Therefore, for high SNR values, e.g., at $SNR = 20$ dB where all four array configurations achieve asymptotic MSE performance, the 2×3 MIMO configuration with external transmitters outperforms all other setups, since it yields the lowest CRB. Conversely, as the CRB of the 2×3 MIMO with adjacent transmitters is only marginally better than that of SIMO, for high SNR values the green MSE curve representing the 2×3 MIMO setup with adjacent transmitters gets closer to the SIMO performance. Finally, the 3×3 MIMO configuration provides a more balanced performance, as it distributes the available transmit power across nine different phase centers.
- As discussed at the beginning of this section, the markers ('×') in Figure 8 denote Monte Carlo simulation results obtained using quasi-orthogonal LFM waveforms. Given the high compression gain of these waveforms, the slight residual cross-talk is negligible. As a result, the simulated points show excellent consistency with the theoretical predictions, which assume perfect orthogonality. This further confirms the effectiveness of the proposed model and approach.

Overall, the proposed array design ensures high estimation accuracy for SNR values equal to or greater than the design SNR, maintaining an estimation accuracy close to the CRB.

To complement this analysis, we conduct an additional comparison at a fixed-pairwise error probability. Specifically, the potential of the 2×3 MIMO with adjacent transmitters and the 3×3 MIMO setups is not fully exploited. Although we ensured a maximum pairwise error probability of 10^{-3} at a design SNR of 15 dB, using array z_1, in these two cases, results in a much lower actual pairwise error probability, namely $5.4 \cdot 10^{-6}$ for the 2×3 MIMO with adjacent transmitter, and $5.2 \cdot 10^{-6}$ for the 3×3 MIMO. This explains the superior performance of these two configurations at the design SNR. However, the constraint was overfulfilled, meaning that with z_1, these setups are already very close to their CRB. Based on this analysis, we conduct a second comparison, assuming a fixed outlier probability, thus selecting a different array from the APP region in each case to ensure the same maximum pairwise error probability. Specifically:

- For SIMO and 2 × 3 MIMO with transmitters at the edges, we keep the array configuration z_1 (one of the few arrays in the APP region for these cases), which achieves a maximum pairwise error probability of about $0.6 \cdot 10^{-3}$.
- For 2 × 3 MIMO with adjacent transmitters, we use array $z_2 = [0\ 2.20\ 3.91]\lambda$.
- For 3 × 3 MIMO, we use array $z_3 = [0\ 1.76\ 3.98]\lambda$.

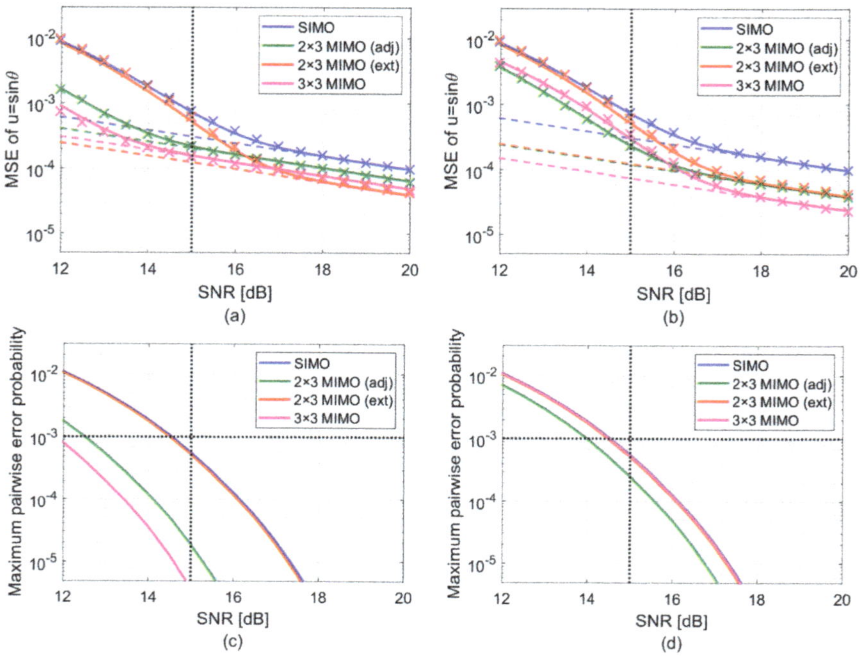

Figure 8. DoA estimation MSE and maximum pairwise error probability as a function of the SNR. The comparisons are performed (**a**,**c**) assuming the same array configuration z_1, achieving different maximum pairwise error probabilities for different SIMO/MIMO configurations, and (**b**,**d**) assuming different array configurations achieving the same maximum pairwise error probability, $P_{max} \approx 0.6 \cdot 10^{-3}$.

Both z_2 and z_3 still achieve $P_{max} \approx 0.6 \cdot 10^{-3}$, bringing them much closer to the constraint $P_{max} \leq 10^{-3}$.

Figure 8b shows the MSE of the different array configurations, averaged within the U_{max} sector, as a function of SNR, while Figure 8d shows the probability of outlier and the maximum pairwise error probability as a function of the SNR. The following observations are in order:

- At the design SNR, all MSE values deviate from their respective CRBs. In such conditions, the 3 × 3 MIMO achieves the best performance due to its lower CRB, while the SIMO exhibits the highest MSE.
- Unlike in Figure 8a, in this case, the threshold region starts at approximately the same SNR for all configurations. Once asymptotic performance is reached, the 3 × 3 MIMO clearly outperforms the 2 × 3 MIMO, which in turn outperforms the SIMO. This improvement is due to the greater design flexibility provided by MIMO configurations compared to SIMO, which allows increasing the global array length, thus achieving a lower CRB.
- As observed in the fixed-array case study, the close correspondence between the theoretical predictions (assuming perfect orthogonality) and the simulation results

(based on quasi-orthogonal LFM waveforms) again demonstrates the robustness of our design approach against small residual cross-talk between the transmitted waveforms.

Overall, the analysis carried out in Figure 8 highlights the twofold advantage of MIMO operation. On one hand, MIMO operation allows increasing the global array length, as it allows us to operate with an effective array with MN elements. In addition to this trivial benefit, MIMO also offers greater design flexibility, allowing us to design more challenging NULA configurations with higher global lengths. The compound effects of these two factors become evident when comparing performance under fixed-array conditions versus fixed maximum pairwise error probability, demonstrating both the structural and design benefits of MIMO in DoA estimation.

Incidentally, we observe that the sparse nature of the designed arrays may impose a lower bound on the operational range due to far-field assumptions. For instance, it is easy to verify that the longest baseline considered (i.e., $D = 3.98\lambda$ for array z_3) corresponds to a Fraunhofer distance of approximately 4 m at $f_c = 2.4$ GHz. This makes this array design best-suited for medium-to-long-range applications, where targets of interest are located at ranges from tens of meters to kilometers. For scenarios requiring operation at very short ranges, the designer could simply add a geometric constraint to the APP region to exclude arrays with excessively large apertures.

To conclude the analysis, in Table 1, we report the average MSEs achieved by different configurations in the fixed-array and fixed-pairwise error probability comparisons.

Table 1. MSEs averaged within U_{max} (in square radians).

Comparison	SNR	SIMO	2 × 3 MIMO Adjacent TX	2 × 3 MIMO External TX	3 × 3 MIMO
Fixed array	15 dB	$6.6 \cdot 10^{-4}$	$1.4 \cdot 10^{-4}$	$5.4 \cdot 10^{-4}$	$0.9 \cdot 10^{-4}$
	20 dB	$0.6 \cdot 10^{-4}$	$0.4 \cdot 10^{-4}$	$0.2 \cdot 10^{-4}$	$0.3 \cdot 10^{-4}$
Fixed pairwise	15 dB	$6.6 \cdot 10^{-4}$	$3.0 \cdot 10^{-4}$	$5.4 \cdot 10^{-4}$	$0.3 \cdot 10^{-4}$
	20 dB	$0.6 \cdot 10^{-4}$	$0.2 \cdot 10^{-4}$	$0.2 \cdot 10^{-4}$	$0.1 \cdot 10^{-4}$

From the table, it is evident that the MSE decreases significantly when MIMO operational modes are used, both at $SNR = 15$ dB and $SNR = 20$ dB. To better illustrate the performance improvement of different MIMO configurations relative to SIMO, Table 2 reports the ratio between the average MSE obtained in the SIMO case and that achieved in each MIMO setup, quantifying the efficiency gains provided by MIMO in terms of estimation accuracy.

Table 2. Ratio between MSEs averaged within U_{max} between SIMO and MIMO.

Comparison	SNR	SIMO	2 × 3 MIMO Adjacent TX	2 × 3 MIMO Edge TX	3 × 3 MIMO
Fixed array	15 dB	1	4.6	1.2	7.3
	20 dB	1	1.3	2.5	2.0
Fixed pairwise	15dB	1	2.2	1.2	2.3
	20 dB	1	3.0	2.5	4.2

Table 2 clearly demonstrates the superior performance of MIMO configurations compared to SIMO across all considered scenarios. Particularly:

- In the fixed-array comparison at SNR = 15 dB, the 3 × 3 MIMO setup achieves the most significant reduction in MSE, with a performance improvement of 7.3 times over SIMO. The 2 × 3 MIMO with adjacent transmitters outperforms the 2 × 3 MIMO with transmitters at the edges, achieving an MSE reduction of 4.6 times compared to SIMO, thanks to its greater robustness to outliers. Finally, the 2 × 3 MIMO with transmitters at the edges offers a more modest improvement of 1.2 times, which is essentially due to its lower CRB, as the outlier probability is not different from that achieved by the SIMO.

- At SNR = 20 dB, all the setups achieve asymptotic performance, so the performance gap between configurations narrows, since the probability of outliers is negligible here. In this case, MIMO remains advantageous with respect to SIMO. Particularly, the 2 × 3 MIMO with transmitters at the edges achieves the highest improvement, showing an MSE 2.5 times lower than that achieved using SIMO.
- In the fixed-pairwise error probability comparison, the performance improvements in the threshold region, namely at SNR = 15 dB, are more similar across the different MIMO setups, since the arrays have been chosen to guarantee a similar probability of outliers across the different setups.
- At SNR = 20 dB, however, the performance improvement increases compared to the fixed-array comparison, since the MIMO setup allowed choosing longer NULA designs. Particularly, the 3 × 3 MIMO significantly outperforms the other configurations, with an MSE improvement factor of 4.2 compared to SIMO, followed by the 2 × 3 MIMO with adjacent transmitters (3.0) and the 2 × 3 MIMO with edge transmitters (2.5).

Overall, these results confirm the dual advantage of MIMO operation. First, MIMO enables an increase in the number of equivalent phase centers, leading to enhanced asymptotic performance. Second, it provides greater design flexibility, improving robustness against outliers and allowing for the implementation of NULA that would not be admissible in a SIMO setup.

After the comparison between SIMO and MIMO setups, we now compare the proposed outlier-controlled NULA design against several standard sparse array configurations. This analysis is performed for the 3 × 3 MIMO case, and the results are shown in Figure 9.

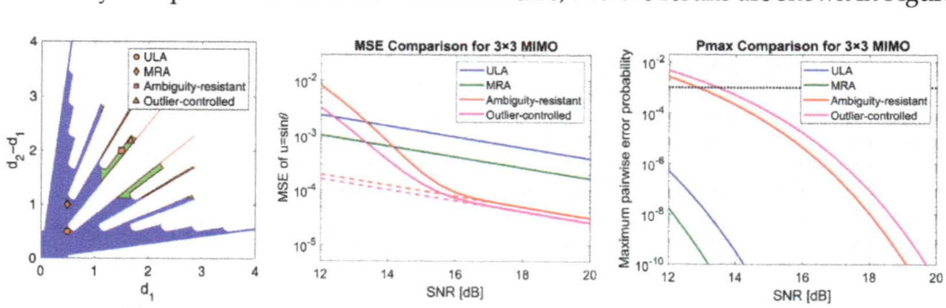

(**a**) (**b**) (**c**)

Figure 9. (**a**) APP region and sparse array designs relative positions in the inter-element distance plane. Comparison of (**b**) DoA estimation MSE and (**c**) maximum pairwise error probability obtained using different sparse array configurations as a function of the SNR.

The first benchmark is the three-element minimum redundancy array (MRA). As demonstrated in [23], the only MRA with $N = 3$ elements is $z_{MRA} = [0, 0.5, 1.5]\lambda$. This configuration yields three unique baseline lengths (1, 2, and 3 in $\lambda/2$ units), thus maximizing the asymptotic angular resolution for a given number of elements. As shown by the '$\Diamond$' marker in Figure 9a, this array is considerably shorter than the array $z_{outlier-controlled} = [0\ 1.67\ 3.85]\lambda$ designed using our strategy ('Δ' marker). Furthermore, we note that this MRA would not be admissible according to the selected minimum inter-element distance constraint ($l \approx 1.12\lambda$), but we include it in the comparison for a comprehensive performance evaluation.

The second benchmark is an ambiguity-resistant NULA designed according to the criteria from [24]. This design strategy requires the two inter-element distances to be coprime multiples of $\lambda/2$. Among the many possible choices, we selected the configuration $z_{ambiguity-resistant} = [0\ 1.5\ 3.5]\lambda$ ('□' marker), which corresponds to coprime spacings of 3 and 4 units of $\lambda/2$. Notably, this array configuration falls within our generated admissible region and satisfies the imposed technological constraints.

Finally, we include a standard $\lambda/2$ ULA, with positions $z_{ULA} = [0,\ 0.5, 1]\lambda$ ('○' marker), as a conservative baseline. This configuration also violates our minimum distance constraint.

Figure 9b,c compare the MSE and maximum pairwise error probability for these configurations. As expected, the short-aperture ULA and MRA achieve asymptotic performance within the considered range of SNR values, exhibiting negligible outlier probability. However, their short baselines result in a relatively high CRB, which limits their asymptotic accuracy.

In contrast, the much sparser outlier-controlled and ambiguity-resistant NULAs achieve a significantly lower CRB, offering superior asymptotic performance. For these arrays, the outlier probability is no longer negligible in the threshold region, but satisfies the required constraint ($P_m < P_{max} = 10^{-3}$, $\forall m$) for SNR values larger than the design SNR of 15 dB.

The ambiguity-resistant design is inherently robust due to its coprime nature, but its maximum pairwise error probability is not explicitly controlled for a specific design SNR. In contrast, our strategy predicts and constrains P_{max} to remain below the design threshold (indicated by the dashed line in Figure 9c) at the target SNR of 15 dB. This controlled robustness, combined with its marginally longer aperture, allows our proposed design to achieve both slightly better asymptotic accuracy and, more importantly, a significantly lower MSE in the critical threshold region.

In conclusion, this comparison highlights the practical value of our proposed strategy. Unlike fixed design criteria, our approach provides a flexible framework that defines an entire region of admissible arrays based on specific design parameters (SNR and P_{max}). This allows a designer to select the optimal configuration for their application, balancing performance trade-offs that are not captured by fixed-design arrays.

5. Conclusions

In this paper, we have extended the outlier-robust NULA design strategy introduced in [29] from SIMO to MIMO radar applications. Specifically, we studied how the number and positioning of transmitters impact both the design procedure and the DoA estimation accuracy. Through theoretical analysis, we have shown that MIMO operation not only enhances DoA estimation accuracy but also enhances array robustness to outliers. This flexibility enables a broader degree of flexibility when designing array configurations for AESA radar.

Moreover, we conducted a comparative analysis between the proposed MIMO-enabled NULA configurations and established NULA designs with inter-element distances at multiples of half wavelength. Our findings highlighted the effectiveness of our design strategy in identifying NULA configurations that achieve higher DoA estimation accuracy. This advantage holds even in scenarios where practical technological constraints make it impossible to resort to conventional design solutions.

The results of our analysis show that a three-element NULA array MIMO radar can be a viable and advantageous solution for low-cost sensors that require reliable and accurate DoA estimation but cannot afford either arrays with a large number of elements or large antenna apertures.

Author Contributions: Conceptualization, A.Q., F.C. and P.L.; Methodology, A.Q., F.C. and P.L.; Software, A.Q.; Validation, A.Q.; Formal analysis, A.Q.; Writing—original draft, A.Q.; Writing—review & editing, F.C. and P.L.; Visualization, A.Q.; Supervision, F.C. and P.L.; Project administration, F.C. and P.L.; Funding acquisition, F.C. and P.L. All authors have read and agreed to the published version of the manuscript.

Funding: This work was partially supported by the Italian National Recovery and Resilience Plan (NRRP) of NextGenerationEU, "Rome Technopole" (CUP B83C22002820006)—Flagship Project 5: "Digital Transition through AESA radar technology, quantum cryptography and quantum communications".

Data Availability Statement: The data presented in this study are available on request from the corresponding author.

Conflicts of Interest: The authors declare no conflict of interest.

Appendix A. 3 × 3 MIMO

As discussed in Section 3, in the 3 × 3 MIMO case, the transmission and reception functionals are equal, since $N = M = 3$, therefore, the likelihood function $L(\phi_1, \phi_2)$ becomes:

$$L(\phi_1, \phi_2) = \left|1 + e^{j\phi_1} + e^{j\phi_2}\right|^4 = [3 + 2cos\phi_1 + 2cos\phi_2 + 2cos(\phi_2 - \phi_1)]^2. \tag{A1}$$

By using the 45° rotated axis system in Equation (15), we can rewrite Equation (A1) as:

$$L(x, y) = \left[1 + 4cosycosx + 4cos^2x\right]^2 \tag{A2}$$

By solving the equation $L(x, y) = L_{max}$ we obtain:

$$cosy = \frac{\sqrt{L_{max}} - 1}{4cosx} - cosx, \tag{A3}$$

yielding:

$$y = \pm acos\left\{\frac{\sqrt{L_{max}} - 1}{4cosx} - cosx\right\} + 2\pi m. \tag{A4}$$

To obtain valid solutions, the argument of the arccosine must be limited as follows:

$$-1 \leq \frac{\sqrt{L_{max}} - 1}{4cosx} - cosx \leq 1. \tag{A5}$$

This constraint is verified by the system of the two inequalities:

$$\begin{cases} \frac{4cos^2x - 4cosx + 1 - \sqrt{L_{max}}}{cosx} \leq 0 \\ \frac{4cos^2x + 4cosx + 1 - \sqrt{L_{max}}}{4cosx} \geq 0 \end{cases}. \tag{A6}$$

Therefore, assuming $L_{max} \geq 1$, the first inequality is verified when:

$$cosx \leq \frac{1 - \sqrt[4]{L_{max}}}{2} \cup 0 \leq cosx \leq \frac{1 + \sqrt[4]{L_{max}}}{2}, \tag{A7}$$

Similarly, the second inequality is verified when:

$$-\frac{1 + \sqrt[4]{L_{max}}}{2} \leq cosx \leq 0 \cup cosx \geq -\frac{1 - \sqrt[4]{L_{max}}}{2}. \tag{A8}$$

Calculating the union of these conditions gives:

$$-\frac{1 + \sqrt[4]{L_{max}}}{2} \leq cosx \leq \frac{1 - \sqrt[4]{L_{max}}}{2} \cup -\frac{1 - \sqrt[4]{L_{max}}}{2} \leq cosx \leq \frac{1 + \sqrt[4]{L_{max}}}{2}, \tag{A9}$$

which results in:

$$-acos\left(\frac{-1+\sqrt[4]{L_{max}}}{2}\right) \le x \le acos\left(\frac{-1+\sqrt[4]{L_{max}}}{2}\right). \tag{A10}$$

Therefore, the locus of points in the (ϕ_1, ϕ_2) plane corresponding to the likelihood function cut at L_{max} is obtained by applying the inverse coordinate transformation to Equation (37) and imposing condition Equation (A10), namely:

$$\begin{cases} \phi_1 = \pm acos\left\{\frac{\frac{L_{max}}{4cos^2x}-1}{4cosx} - cosx\right\} + 2\pi h - x \\ \phi_2 = \pm acos\left\{\frac{\frac{L_{max}}{4cos^2x}-1}{4cosx} - cosx\right\} + 2\pi k + x \end{cases} \tag{A11}$$

within the limits of Equation (A10) for x.

Appendix B. 2 × 3 MIMO

In the 2 × 3 MIMO case, with the phase reference chosen at the receiving-only array element, the transmission functional includes only the phase terms ϕ_1 and ϕ_2

$$\Psi_2(\phi_1, \phi_2) = \left|e^{j\phi_1} + e^{j\phi_2}\right|^2 = 2 + 2cos(\phi_2 - \phi_1), \tag{A12}$$

while the reception functional is still the square root of Equation (A1) $\Psi_3(\phi_1, \phi_2)$. Therefore, the likelihood function $L(\phi_1, \phi_2)$ is:

$$L(\phi_1, \phi_2) = 2[1 + cos(\phi_2 - \phi_1)][3 + 2cos\phi_1 + 2cos\phi_2 + 2cos(\phi_2 - \phi_1)]. \tag{A13}$$

By using the 45° rotated axis system in Equation (15), we can rewrite Equation (A13) as:

$$L(x, y) = 4cos^2x\left(1 + 4cosycosx + 4cos^2x\right) \le L_{max} \tag{A14}$$

The solution of the associated equation is:

$$cosy = \frac{\frac{L_{max}}{4cos^2x} - 1}{4cosx} - cosx, \tag{A15}$$

which yields

$$y = \pm acos\left\{\frac{\frac{L_{max}}{4cos^2x} - 1}{4cosx} - cosx\right\} + 2\pi m. \tag{A16}$$

To obtain valid solutions, the following constraint must be imposed on the arccosine argument:

$$-1 \le \frac{\frac{L_{max}}{4cos^2x} - 1}{4cosx} - cosx \le 1, \tag{A17}$$

This constraint is verified by the system of the two inequalities:

$$\begin{cases} \frac{cos^4x - cos^3x + \frac{cos^2x}{4} - \frac{L_{max}}{16}}{cos^3x} \le 0 \\ \frac{cos^4x + cos^3x + \frac{cos^2x}{4} - \frac{L_{max}}{16}}{cos^3x} \ge 0 \end{cases}. \tag{A18}$$

The numerator of the first inequality has only two real roots:

$$R_{1,2} = \frac{1 \pm \sqrt{1 + 4\sqrt{L_{max}}}}{4} \in \mathbb{R}. \tag{A19}$$

Therefore, assuming $L_{max} \geq 1$, the first inequality is negative if

$$cosx \leq \frac{1-\sqrt{1+4\sqrt{L_{max}}}}{4} \cup 0 \leq cosx \leq \frac{1+\sqrt{1+4\sqrt{L_{max}}}}{4}. \quad \text{(A20)}$$

Similarly, the numerator of the second inequality has only two real roots:

$$R_{3,4} = -\frac{1 \pm \sqrt{1+4\sqrt{L_{max}}}}{4} \in \mathbb{R}. \quad \text{(A21)}$$

Therefore, assuming $L_{max} \geq 1$, the second inequality is positive if

$$-\frac{1+\sqrt{1+4\sqrt{L_{max}}}}{4} \leq cosx \leq 0 \cup cosx \geq -\frac{1-\sqrt{1+4\sqrt{L_{max}}}}{4}. \quad \text{(A22)}$$

The union of conditions Equations (A20) and (A22) yields:

$$\begin{array}{c} -\frac{1+\sqrt{1+4\sqrt{L_{max}}}}{4} \leq cosx \leq \frac{1-\sqrt{1+4\sqrt{L_{max}}}}{4} \\ \cup \\ -\frac{1-\sqrt{1+4\sqrt{L_{max}}}}{4} \leq cosx \leq \frac{1+\sqrt{1+4\sqrt{L_{max}}}}{4}, \end{array} \quad \text{(A23)}$$

which results in:

$$-acos\left(\frac{-1+\sqrt{1+4\sqrt{L_{max}}}}{4}\right) \leq x \leq acos\left(\frac{-1+\sqrt{1+4\sqrt{L_{max}}}}{4}\right). \quad \text{(A24)}$$

Therefore, the locus of points in the (ϕ_1, ϕ_2) plane corresponding to the likelihood function cut at L_{max} is obtained by applying the inverse coordinate transformation to equation Equation (A16) and imposing condition Equation (A24), namely:

$$\begin{cases} \phi_1 = \pm acos\left\{\frac{\frac{L_{max}}{4cos^2x}-1}{4cosx} - cosx\right\} + 2\pi h - x \\ \phi_2 = \pm acos\left\{\frac{\frac{L_{max}}{4cos^2x}-1}{4cosx} - cosx\right\} + 2\pi k + x \end{cases}, \quad \text{(A25)}$$

within the limits of Equation (A24) for x.

$$-acos\left(\frac{-1+\sqrt{1+4\sqrt{L_{max}}}}{4}\right) \leq x \leq acos\left(\frac{-1+\sqrt{1+4\sqrt{L_{max}}}}{4}\right). \quad \text{(A26)}$$

References

1. Webster, A.R.; Jones, J.; Ellis, K.J. Direction finding using a linear three-element interferometer approach. *Radio Sci.* **2002**, *37*, 1095. [CrossRef]
2. Tu, Y.; Feng, B.; Wang, M.; He, X.; Zeng, Q. A three-element dual-polarized antenna array with beamwidth reconfiguration. In Proceedings of the 2017 Sixth Asia-Pacific Conference on Antennas and Propagation (APCAP), Xi'an, China, 16–19 October 2017.
3. Lammert, V.; Hamouda, M.; Weigel, R.; Issakov, V. A 122 GHz On-Chip 3-Element Patch Antenna Array with 10 GHz Bandwidth. In Proceedings of the 2020 50th European Microwave Conference (EuMC), Utrecht, the Netherlands, 12–14 January 2021.
4. Pfahler, T.; Vossiek, M.; Schür, J. Compact and Broadband 300 GHz Three-Element on-chip Patch Antenna. In Proceedings of the 2023 IEEE Radio and Wireless Symposium (RWS), Las Vegas, NV, USA, 22–25 January 2023.
5. Cheng, Y.F.; Cheng, K.K.M. A Novel and Simple Decoupling Method for a Three-Element Antenna Array. *IEEE Antennas Wirel. Propag. Lett.* **2017**, *16*, 1072–1075. [CrossRef]
6. Apostolov, P.; Yurukov, B.; Stefanov, A. Efficient Three-Element Binomial Array Antenna. In Proceedings of the 2019 PhotonIcs & Electromagnetics Research Symposium—Spring (PIERS-Spring), Rome, Italy, 17–20 June 2019.
7. Filippini, F.; Martelli, T.; Colone, F.; Cardinali, R. Target DoA estimation in passive radar using non-uniform linear arrays and multiple frequency channels. In Proceedings of the 2018 IEEE Radar Conference, Oklahoma, OK, USA, 23–27 April 2018.

8. van Trees, H.L. *Detection, Estimation, and Modulation Theory*; Wiley: New York, NY, USA, 1968; Part I; ISBN 978-0471095170.
9. Richmond, C.D. Mean-squared error and threshold SNR prediction of maximum-likelihood signal parameter estimation with estimated colored noise covariances. *IEEE Trans. Inf. Theory* **2006**, *52*, 2146–2164. [CrossRef]
10. Barankin, E.W. Locally Best Unbiased Estimates. *Ann. Math. Stat.* **1949**, *20*, 477–501. [CrossRef]
11. Stoica, P.; Nehorai, A. MUSIC, maximum likelihood, and Cramer-Rao bound. *IEEE Trans. Acoust. Speech Signal Process.* **1989**, *37*, 720–741. [CrossRef]
12. Stoica, P.; Nehorai, A. Performance study of conditional and unconditional direction-of-arrival estimation. *IEEE Trans. Acoust. Speech Signal Process.* **1990**, *38*, 1783–1795. [CrossRef]
13. Rife, D.C.; Boorstyn, R.B. Single-tone parameter estimation from discrete-time observations. *IEEE Trans. Inf. Theory* **1974**, *20*, 591–598. [CrossRef]
14. Ziv, J.; Zakai, M. Some lower bounds on signal parameter estimation. *IEEE Trans. Inf. Theory* **1969**, *15*, 386–391. [CrossRef]
15. Bell, K.L.; Ephraim, Y.; van Trees, H.L. Explicit Ziv-Zakai lower bound for bearing estimation. *IEEE Trans. Signal Process.* **1996**, *44*, 2810–2824. [CrossRef]
16. Athley, F. Threshold region performance of maximum likelihood direction of arrival estimators. *IEEE Trans. Signal Process.* **2005**, *53*, 1359–1373. [CrossRef]
17. Filippini, F.; Colone, F.; De Maio, A. Threshold region performance of multicarrier maximum likelihood direction of arrival estimators. *IEEE Trans. Aerosp. Electron. Syst.* **2019**, *55*, 3517–3530. [CrossRef]
18. Schmidt, R. Multiple emitter location and signal parameter estimation. *IEEE Trans. Antennas Propag.* **1986**, *34*, 276–280. [CrossRef]
19. Roy, R.; Kailath, T. ESPRIT-estimation of signal parameters via rotational invariance techniques. *IEEE Trans. Acoust. Speech Signal Process.* **1989**, *37*, 984–995. [CrossRef]
20. Donoho, D.L. Compressed sensing. *IEEE Trans. Inf. Theory* **2006**, *52*, 1289–1306. [CrossRef]
21. Candès, E.J.; Romberg, J.; Tao, T. Robust uncertainty principles: Exact signal reconstruction from highly incomplete frequency information. *IEEE Trans. Inf. Theory* **2006**, *52*, 489–509. [CrossRef]
22. Bazzi, A.; Slock, D.T.M.; Meilhac, L. Sparse recovery using an iterative Variational Bayes algorithm and application to AoA estimation. In Proceedings of the 2016 IEEE International Symposium on Signal Processing and Information Technology (ISSPIT), Limassol, Cyprus, 12–14 December 2016; pp. 197–202. [CrossRef]
23. Moffet, A. Minimum-redundancy linear arrays. *IEEE Trans. Antennas Propag.* **1968**, *16*, 172–175. [CrossRef]
24. Goodwin, R.L. *Ambiguity-Resistant Three-and Four-Channel Interferometers*; NRL Report 8005; Naval Research Laboratory: Washington, DC, USA, 1976.
25. Athley, F. Optimization of element positions for direction finding with sparse arrays. In Proceedings of the 11th IEEE Signal Processing Workshop Statistical Signal Process, Singapore, 8 August 2001.
26. Lee, J.H.; Lee, J.H.; Woo, J.M. Method for Obtaining Three- and Four-Element Array Spacing for Interferometer Direction-Finding System. *IEEE Antennas Wirel. Propag. Lett.* **2016**, *15*, 897–900. [CrossRef]
27. Lee, J.H.; Kim, J.K.; Ryu, H.K.; Park, Y.J. Multiple Array Spacings for an Interferometer Direction Finder With High Direction-Finding Accuracy in a Wide Range of Frequencies. *IEEE Antennas Wirel. Propag. Lett.* **2018**, *17*, 563–566. [CrossRef]
28. Horng, W.Y. An Efficient DOA Algorithm for Phase Interferometers. *IEEE Trans. Aerosp. Electron. Syst.* **2020**, *56*, 1819–1828. [CrossRef]
29. Quirini, A.; Filippini, F.; Bongioanni, C.; Colone, F.; Lombardo, P. A Flexible Design Strategy for Three-Element Non-Uniform Linear Arrays. *Sensors* **2023**, *23*, 4872. [CrossRef] [PubMed]
30. Proakis, J.G. *Digital Communications*, 4th ed.; McGraw-Hill: New York, NY, USA, 2001; ISBN 978-0071181839.

sensors

MDPI

Article

Direct Position Determination of Wideband Source over Multipath Environment: Combining Taylor Expansion and Subspace Data Fusion in the Cross-Spectrum Domain

Heng Chai [1], Xinjian Yin [1,*], Hao Hu [1,2] and Xiaofei Zhang [1]

[1] College of Electronic Information Engineering, Nanjing University of Aeronautics and Astronautics, Nanjing 211106, China; chaiheng8888@163.com (H.C.); huhao2190@nuaa.edu.cn (H.H.); zhangxiaofei@nuaa.edu.cn (X.Z.)

[2] Jiangsu Automation Research Institute, Lianyungang 222000, China

* Correspondence: xinjianyin@nuaa.edu.cn

Abstract: Position localization of wideband source over multipath environment is addressed in this paper. Traditional methods generally estimate intermediate parameters first and then use these parameters to construct equations for determining the source position. However, the localization accuracy of such methods deteriorates significantly in the presence of multipath effects. In this paper, a direct position determination method combining Taylor expansion and subspace data fusion in the cross-spectrum domain is proposed. The method constructs the data model based on the cross-spectrum of the received signals from arbitrary sensor pairs, effectively avoiding the loss of the available information. Subsequently, forward spatial smoothing is used to address the rank-deficiency problem caused by the multipath effect. Finally, a cost function using subspace data fusion is constructed, and the optimal value is derived via first-order Taylor expansion to compensate for the position estimation bias. The proposed method shows higher localization accuracy compared to state-of-the-art methods. The numerical and experimental results validate the superior localization performance of the proposed algorithm.

Keywords: direct position determination; distributed sensors; multipath environment; Taylor expansion

1. Introduction

Emitter localization techniques, which rely on the distributed passive stations, have significantly contributed to a multitude of fields, including radar and sonar systems, wireless communication, automotive technologies, and signal processing [1,2]. The traditional two-step algorithm [3] is a commonly used position determination method in the early days that contains two independent steps. This method first obtains position-related parameters using the received signals from the observation stations, including the direction of arrival (DOA) [4], time of arrival (TOA) [5], time difference of arrival (TDOA) [6,7], received signal strength (RSS) [8], and frequency difference of arrival (FDOA), etc. Subsequently, the emitter's position is determined by formulating corresponding equations with the previously obtained parameters. Although the two-step method is widely used, it is still not optimal. The most important reason is that it ignores the correlation between the received signals from the same emitter.

Recently, with the continuous development and improvement of the Direct Positioning Determination (DPD) algorithm [9], this problem has been resolved. In [10], Weiss was the first to systematically introduce this concept and demonstrated its superiority over two-step positioning methods under low SNR conditions. The algorithm employs the Least Square (LS) criterion and Maximum Likelihood (ML) criterion to address the localization problems under unknown noise and known Gaussian noise conditions, respectively. The author in [11] employs a single moving array to collect signals and introduces a novel Subspace Data Fusion (SDF) method, which can be regarded as a generalized form of the MUSIC algorithm [12]. This method significantly improves localization performance and enables the localization of multiple emitters, but requires determining the number of sources in advance. The DPD algorithm described in [13], which leverages the Minimum Variance Distortionless Response (MVDR) approach, offers high-resolution estimation even when dealing with multiple emitters whose numbers are unknown. In [14], the author proposed a DPD algorithm for interference sources in satellite systems based on a propagation operator. This approach reduces the computational complexity of subspace calculations compared with the MUSIC-based methods. In [15], Quantum-behaved Particle Swarm Optimization (QPSO) is applied to determine the optimal array response, thus proposing a Multi-array Data Fusion (MDF)-based technique. In [16], the author proposed a DPD method based on the signal-to-noise ratio (SNR) weighting and the propagation operator to address the issue of varying received SNRs at different observation stations. The author in [17] proposed a closed-form DPD algorithm for linear emitters based on phase alignment and optimal weighted least square (OWLS), referred to as the PAOWLS method.

The aforementioned DPD methods mainly rely on array stations, resulting in significant hardware expenses. In contrast, the DPD method based on TDOA can be implemented using distributed single sensors, yet there is a scarcity of algorithms currently. In [18], the TDOA-based DPD algorithm derived by the maximum likelihood estimator (MLE) is proposed, which provides the Fisher Information Matrix (FIM) in the TDOA scenario, and the author derives the Cramér–Rao Bound (CRB). In [19], the author offers a DPD algorithm for multiple unknown signal emitters based on Hough transform (HT). The author in [20] gives a determinant-based method that uses matrix decomposition and the orthogonal relationship between the noise subspace and the signal manifold to obtain the spectral function. The method delivers superior positioning accuracy for multi-target localization, while the high dimension of the fusion matrix increases computational complexity. Signal time-domain data segmentation is used in [21] to reduce the computational complexity of matrix decomposition. Subsequently, a spectrum function is developed via multiple-frequency function fusion (MFF) to determine the position. In [22], the author introduces a DPD method based on Parallel Factor (PARAFAC), which acquires the time-delay matrix through rapid iteration and then constructs the cost function, providing enhanced performance and reduced complexity. In [23], the authors propose a subspace focusing algorithm for multi-array data fusion, which improves the positioning performance of wideband sources using distributed arrays. In addition, there are still many technologies [24–36] that have made outstanding contributions to positioning. Most of the aforementioned approaches just consider ideal environments, but only a few algorithms consider localization in multipath environments.

In this paper, a direct position determination algorithm based on TDOA under the multipath environment is proposed. The proposed algorithm jointly employs the cross-spectrum and Taylor compensation, which is proficient in efficiently resolving challenges in multipath environments. The following are the main contributions of this work:

1. We construct the data model based on the cross-spectrum between received signals from any pair of sensors. This method fully extracts the position information from the received signals, thereby avoiding any loss of available information, and can be effectively applied to high-resolution models. Moreover, we further use forward spa-

tial smoothing to address the issue of covariance matrix rank deficiency in multipath environments.

2. We offer a cost function based on forward spatial smooth and subspace data fusion to obtain the initial estimation, and then first-order Taylor expansion is employed to obtain off-grid compensation on the initial estimation, enhancing the localization performance for emitters that do not lie on the search grid.
3. We have evaluated the performance of our algorithm across multiple circumstances using comprehensive simulation studies and actual experiments. The results of these simulations demonstrate our algorithm's remarkable efficacy and adaptability, underscoring its significant potential for application in multipath environments.

The structure of this paper is as follows: Section 2 describes the model of the received signals. In Section 3, cross-spectrum data model is introduced. Then, Section 4 introduces the proposed algorithm. In Section 5, the proposed algorithm is compared with other algorithms through numerical simulation. Then, the real-world experiment is conducted in Section 6 to verify the viability of the proposed algorithm. Finally, Section 7 concludes the paper.

Notations: Bold characters indicate vectors (e.g., $\mathbf{a}$), bold italic characters indicate matrices (e.g., $\boldsymbol{A}$). $[\bullet]^T$, $[\bullet]^*$, $[\bullet]^H$ and $[\bullet]^+$ stand for transpose, conjugation, conjugate transpose and Moore–Penrose inverse, respectively. $[\bullet]_n$ denote the n-th element of the vector.

2. Signal Model

Consider that there are L observation sensors and the l-th sensor located at $\mathbf{S}_l = [X_l, Y_l]^T$. We suppose that all sensors are time-synchronized. Suppose that there is an unknown far-field emitter that is confined to a plane with a fixed altitude, and it transmits a wide-band signal. Assuming the unknown emitter is located in $\mathbf{p} = [x_p, y_p]^T$. The localization scenario within a multipath environment is illustrated in Figure 1.

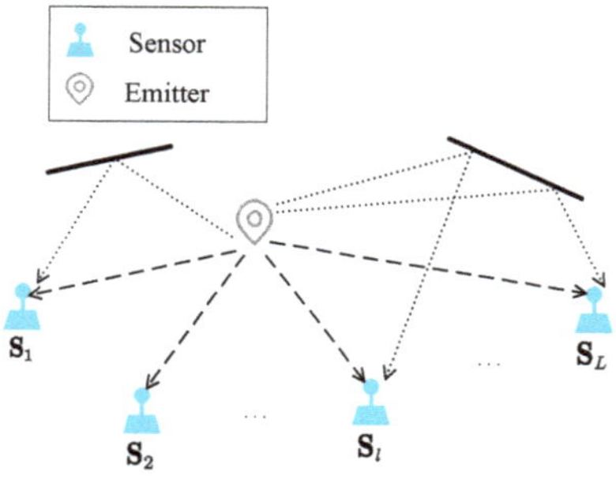

Figure 1. Location scenario in a multipath environment.

Then, the received signal of the l-th sensor can be expressed as

$$x_l(n) = \sum_{m=1}^{M} \lambda_{lm} s(n - \tau_{lm}) + \sigma_l(n), n = 1, 2, \ldots, N_a, \quad (1)$$

where M is the number of multipath components, N_a is the number of samples, λ_{lm} is the attenuation coefficient caused by the multipath environment, $s(n)$ is the unknown wideband signal transmitted by the emitter, $\sigma_l(n)$ is a white, zero-mean, complex, additive Gaussian noise that is independent of the signal, and τ_{lm} is the time delay from the emitter to the l-th sensor in the m-th path.

Considering that the number of sampling points N_a is typically large to ensure accurate information capture, to reduce computational complexity and to facilitate the construction of multi-snapshot data, we segment the received data into smaller segments of length $N = N_a/K$, where K denotes the total number of segments. Then, the k-th part of the received signal can be denoted as

$$x_{lk}(n_k) = \sum_{m=1}^{M} \lambda_{lm} s(n_k - \tau_{lm}) + \sigma_{lk}(n_k), n_k = 1 + (k-1)N, 2 + (k-1)N, \ldots, kN. \quad (2)$$

It should be noted that, although the signal has been segmented, the correlation between these segments remains unchanged.

To extract the time-delay information from the received signal, the cross-correlation function between the signals received by each sensor and the reference signal is typically computed. In this paper, we derive the cross-correlation between any two sensors to fully exploit all the available information. From (2), the k-th part's signal from the i-th and j-th sensor can be expressed as

$$\begin{cases} x_{ik}(n_k) = \sum_{p=1}^{P} \lambda_{ip} s(n_k - \tau_{ip}) + \sigma_{ik}(n_k) \\ x_{jk}(n_k) = \sum_{q=1}^{Q} \lambda_{jq} s(n_k - \tau_{jq}) + \sigma_{jk}(n_k), \end{cases} \quad (3)$$

where P, Q is the multipath number.

3. Pre-Procesing in the Cross-Spectrum Domain

The cross-correlation function can be denoted as

$$R_{ijk}(\tau) = E[x_{ik}(n_k) x_{jk}^H(n_k + \tau)], \quad (4)$$

where $E(\cdot)$ represents expectation. Considering that signal is independent of the noise, by substituting Equation (3) into Equation (4), it can be rewritten as

$$\begin{aligned} R_{ijk}(\tau) &= E\left[\sum_{p=1}^{P} \lambda_{ip} s(n_k - \tau_{ip}) \sum_{q=1}^{Q} \lambda_{jq}^H s^H(n_k - \tau_{jq} + \tau)\right] \\ &= \sum_{p=1}^{P} \sum_{q=1}^{Q} \lambda_{ip} \lambda_{jq}^H E\left[s(n_k - \tau_{ip}) s^H(n_k - \tau_{jq} + \tau)\right] \\ &= \sum_{p=1}^{P} \sum_{q=1}^{Q} \lambda_{ip} \lambda_{jq}^H R_{s_k}(\tau - (\tau_{jq} - \tau_{ip})) \\ &= \sum_{p=1}^{P} \sum_{q=1}^{Q} \lambda_{ip} \lambda_{jq}^H R_{s_k}(\tau - \Delta\tau_{ijpq}), \end{aligned} \quad (5)$$

where $R_{s_k}(\tau)$ is the self-correlation function of $s(n_k)$ and $\Delta\tau_{ijpq} = \tau_{jq} - \tau_{ip}$. Considering that $R_{ijk}(\tau)$ and $R_{jik}(\tau)$ convey identical information, there are $L(L-1)/2$ unique cross-correlation pairs among the L sensors.

By applying discrete Fourier transform (DFT) on $R_{hk}(\tau)$, the corresponding cross-spectrum is obtained as

$$G_{ijk}(f) = G_{s_k}(f) \sum_{p=1}^{P} \sum_{q=1}^{Q} \lambda_{ip} \lambda_{jq}^H e^{-j2\pi f \Delta\tau_{ijpq}}, \quad (6)$$

where $G_{s_k}(f)$ is the self-spectrum of $s_k(n_k)$.

To utilize the time-delay information for the subsequent processing of the received signal, we assume that the selected reference signal is free of multipath components. In practice, a signal with prior information is typically selected as the reference signal. Then, the reference signal can be written as

$$x_{0k}(n_k) = \lambda_0 s(n_k - \tau_0) + \sigma_{0k}(n_k), \tag{7}$$

where λ_{lm} is the attenuation coefficient.

Then, the self-correlation of $x_{0k}(n_k)$ can be denoted as

$$\begin{aligned} R_{x_{0k}}(\tau) &= E[x_{0k}(n_k) x_{0k}^H(n_k + \tau)] \\ &= E\left[\lambda_0 s(n_k - \tau_0) \lambda_0^H s^H(n_k - \tau_0 + \tau)\right] \\ &+ E[\sigma_{0k}(n_k) \sigma_{0k}(n_k + \tau)] \\ &= \lambda_0 \lambda_0^H R_s(\tau) + R_{\sigma_{0k}}(\tau), \end{aligned} \tag{8}$$

so the self-spectrum can be written as

$$G_{x_{0k}}(f) = \lambda_0 \lambda_0^H G_{s_k}(f) + G_{\sigma_{0k}}(f). \tag{9}$$

Then, Equation (6) can be rewritten as

$$G_{ijk}(f) = (G_{x_{0k}}(f) - G_{\sigma_{0k}}(f)) \sum_{p=1}^{P} \sum_{q=1}^{Q} \frac{\lambda_{ip} \lambda_{jq}^H}{\lambda_0 \lambda_0^H} e^{-j2\pi f \Delta\tau_{ijpq}}. \tag{10}$$

Thus, the TDOA information contained in the received signal has been extracted in the form of the cross-spectrum. Equation (10) can be rewritten as

$$g_{ijk}(f) = \frac{G_{ijk}(f)}{G_{x_{0k}}(f)} = \sum_{p=1}^{P} \sum_{q=1}^{Q} \mu_{ijpq} e^{-j2\pi f \Delta\tau_{ijpq}} + \gamma_{ijk}(f) \tag{11}$$

$$\gamma_{ijk}(f) = -\frac{G_{\sigma_{0k}}(f)}{G_{x_{0k}}(f)} \sum_{p=1}^{P} \sum_{q=1}^{Q} \mu_{ijpq} e^{-j2\pi f \Delta\tau_{ijpq}}, \tag{12}$$

where $\mu_{ijpq} = \frac{\lambda_{ip} \lambda_{jq}^H}{\lambda_0 \lambda_0^H}$, $\gamma(f)$ can be regarded as the noise term that is independent of $G_{x_{1k}}(f)$ and follows a Gaussian distribution.

In order to use the high-resolution spectrum estimation method, we would like to sample $g_{ijk}(f)$ uniformly in the frequency domain to obtain the effective vectors. Let ω_0 denote the first chosen frequency point, and we select the valid portion of the cross-spectrum with N_f points, then we get

$$\boldsymbol{G}_{ijk} = \boldsymbol{A}_{ij} \boldsymbol{\mu}_{ij} + \boldsymbol{\gamma}_{ijk} \in \mathbb{C}^{N_f \times 1}, \tag{13}$$

where

$$
\begin{aligned}
G_{ijk} &= \left[g_{ijk}(f_0) \quad g_{ijk}(f_1) \quad \dots \quad g_{ijk}(f_{N_f-1}) \right]^T \\
A_{ij} &= \left[\mathbf{a}_{ij11}(\mathbf{p}) \quad \dots \quad \mathbf{a}_{ijpq}(\mathbf{p}) \quad \dots \quad \mathbf{a}_{ijPQ}(\mathbf{p}) \right] \\
\mathbf{a}_{ijpq}(\mathbf{p}) &= \left[e^{-j2\pi f_0 \Delta\tau_{ijpq}} \quad e^{-j2\pi f_1 \Delta\tau_{ijpq}} \quad \dots \quad e^{-j2\pi f_{N_f-1} \Delta\tau_{ijpq}} \right]^T \\
\mu_{ij} &= \left[\mu_{ij11} \quad \dots \quad \mu_{ijpq} \quad \dots \quad \mu_{ijPQ} \right]^T \\
\gamma_{ijk} &= \left[\gamma_{ijk}(f_0) \quad \gamma_{ijk}(f_1) \quad \dots \quad \gamma_{ijk}(f_{N_f-1}) \right]^T,
\end{aligned}
\tag{14}
$$

where $[\bullet]^T$ is transpose, A_{ij} is the signal manifold matrix, and $a_{ijpq}(\mathbf{p})$ is the signal vector. Combining all data, we can get

$$
G_{ij} = \left[G_{ij1} \quad G_{ij2} \quad \dots \quad G_{ijK} \right]. \tag{15}
$$

4. The Proposed Algorithm

In this section, based on the cross-spectrum model presented in Section 3, a DPD algorithm is proposed. We call it "DPD-JTAC" as it is based on all cross-spectrum information and Taylor compensation.

4.1. Forward Spatial Smoothing Process

Then, the covariance matrix of the ij-th cross-spectrum vector can be denoted as

$$
\hat{R}_{ij} = \frac{1}{K} G_{ij} G_{ij}^H = A_{ij} \hat{R}_{S,ij} A_{ij} + \hat{R}_{W,ij}, \tag{16}
$$

where $\hat{R}_{S,ij} = E[\mu_{ij}\mu_{ij}^H]$ and $\hat{R}_{W,ij} = E[\gamma_{ijk}\gamma_{ijk}^H]$.

However, in a multipath environment, $R_{S,ij}$ may not be nonsingular, which means that directly using R_{ij} for high-resolution spectrum estimation is not feasible. Therefore, we subsequently employ the forward spatial smoothing [37], as depicted in Figure 2, to solve this problem. Let $g_{ijkd}(d = 1, 2, \dots, D) \in \mathbb{C}^{N_f \times 1}$ denote the d-th observation vector of $g_{ijk}(f)$ and it can be expressed as

$$
g_{ijkd} = \left[g_{ijk}(f_{d-1}) \ g_{ijk}(f_{d-1+D})] \ \dots \ g_{ijk}(f_{d-1+D(N_f-1)}) \right]^T, \tag{17}
$$

where D denotes the sampling interval. Thus, the covariance matrix of the smoothed matrix can be given by

$$
\tilde{R}_{ij} = \frac{1}{K} \sum_{k=1}^{K} \mathcal{G}_{ijk} \mathcal{G}_{ijk}^H, \tag{18}
$$

where $\mathcal{G}_{ijk} = \left[g_{ijk1} \quad g_{ijk2} \quad \dots \quad g_{ijkD} \right]$.

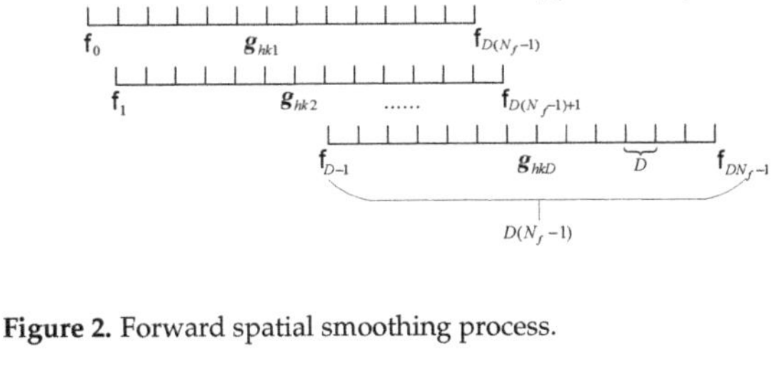

Figure 2. Forward spatial smoothing process.

4.2. Subspace Data Fusion

Then, the eigen-decomposition of $\tilde{\boldsymbol{R}}_{ij}$ is performed to obtain the subspace, which can be expressed as

$$\tilde{\boldsymbol{R}}_{ij} = \boldsymbol{U}_s(ij)\boldsymbol{\Lambda}_s(ij)\boldsymbol{U}_s^H(ij) + \boldsymbol{U}_n(ij)\boldsymbol{\Lambda}_n(ij)\boldsymbol{U}_n^H(ij), \tag{19}$$

where $\boldsymbol{\Lambda}_s(ij)$ and $\boldsymbol{\Lambda}_n(ij)$ are the diagonal matrices containing the largest eigenvalue and the remaining $N_f - 1$ eigenvalues, respectively. And $\boldsymbol{U}_s(ij)$ and $\boldsymbol{U}_n(ij)$ represent the signal subspace and noise subspace, respectively, spanned by the corresponding eigenvectors. According to [10], we have the orthogonal relationship in noise subspace and signal vector, thus, we can get

$$\mathbf{a}_{ijpq}^H(\mathbf{p})\boldsymbol{U}_n(ij) = \mathbf{0}, \tag{20}$$

where $\mathbf{0}$ denotes the zero vector. Since we only need to determine the true position of the source corresponding to the main path, defining $\tilde{\mathbf{a}}_{ij}(\mathbf{p}) = \mathbf{a}_{ij11}(\mathbf{p})$, Equation (20) can be rewritten as:

$$\tilde{\mathbf{a}}_{ij}^H(\mathbf{p})\boldsymbol{U}_n(ij) = \mathbf{0}. \tag{21}$$

However, due to the influence of noise in the received signal, the above equation can only approximately achieve an orthogonal relationship. Therefore, the cost function at any point $\mathbf{p}_0 = [x, y]^T$ in the interested area can be denoted as

$$F(\mathbf{p}_0) = \sum_{i=1}^{L-1}\sum_{j=i+1}^{L} \frac{1}{\tilde{\mathbf{a}}_{ij}^H(\mathbf{p}_0)\boldsymbol{U}_n(ij)\boldsymbol{U}_{ij}^H(ij)\tilde{\mathbf{a}}_{ij}(\mathbf{p}_0)}, \tag{22}$$

where

$$\tilde{\mathbf{a}}_{ij}(\mathbf{p}_0) = \left[e^{-j2\pi f_0 \Delta\tau_{ij11}} \;\; e^{-j2\pi f_D \Delta\tau_{ij11}} \;\; \ldots \;\; e^{-j2\pi f_{D(N_f-1)}\Delta\tau_{ij11}}\right]^T \tag{23}$$

$$\Delta\tau_{ij11} = \frac{\|\mathbf{S}_i - \mathbf{p}_0\|}{c} - \frac{\|\mathbf{S}_j - \mathbf{p}_0\|}{c}. \tag{24}$$

The initial position of the emitter can be determined by searching the spectral peak of the cost function over the meshed region.

4.3. Taylor Expansion

The search-based algorithm is generally constrained by the resolution of the search grid, and the position of the emitter may not coincide with the grid point, potentially resulting in biased estimation. To improve the localization accuracy, an off-grid compensation based on the Taylor formula is applied to the initial estimate.

Let $\mathbf{p}_{ini} = [\hat{x}_{ini}, \hat{y}_{ini}]^T$ denote the initial position, in order to use the all cross-spectrum information, we define

$$\boldsymbol{b}(\mathbf{p}) = \begin{bmatrix} \tilde{\mathbf{a}}_{11}(\mathbf{p}) \\ \vdots \\ \tilde{\mathbf{a}}_{ij}(\mathbf{p}) \\ \vdots \\ \tilde{\mathbf{a}}_{(L-1)L}(\mathbf{p}) \end{bmatrix}. \tag{25}$$

$$\boldsymbol{U}_n = \begin{bmatrix} \boldsymbol{U}_n(11) \\ \vdots \\ \boldsymbol{U}_n(ij) \\ \vdots \\ \boldsymbol{U}_n((L-1)L) \end{bmatrix}. \tag{26}$$

Due to the orthogonal relationship in noise subspace and signal vector, we have

$$\boldsymbol{U}_n^H \boldsymbol{b}(\mathbf{p}) = \mathbf{0}. \tag{27}$$

Then, perform a first-order Taylor expansion of Equation (25) at $\mathbf{p} = \mathbf{p}_{ini}$

$$\boldsymbol{U}_n^H(\boldsymbol{b}(\mathbf{p}_{ini}) + \frac{\partial \boldsymbol{b}(\mathbf{p}_{ini})}{\partial \hat{x}_{ini}}(x - \hat{x}_{ini}) + \frac{\partial \boldsymbol{b}(\mathbf{p}_{ini})}{\partial \hat{y}_{ini}}(y - \hat{y}_{ini})) = 0. \tag{28}$$

Defining the deviation term as $\Delta x = x - \hat{x}_{ini}$, $\Delta y = y - \hat{y}_{ini}$, the least square solution of the deviation term can be denoted as

$$\begin{bmatrix} \Delta x \\ \Delta y \end{bmatrix} = Re\left[-\left(\boldsymbol{U}_n^H\begin{bmatrix} \frac{\partial \boldsymbol{b}(\mathbf{p}_{ini})}{\hat{x}_{ini}} & \frac{\partial \boldsymbol{b}(\mathbf{p}_{ini})}{\hat{y}_{ini}} \end{bmatrix}\right)^+ \boldsymbol{U}_n^H \boldsymbol{b}(\mathbf{p}_{ini})\right] \tag{29}$$

According to (25), we can derive that

$$\frac{\partial \boldsymbol{b}(\mathbf{p}_{ini})}{\partial \hat{x}_{ini}} = \begin{bmatrix} \frac{\partial \tilde{\mathbf{a}}_{11}^T(\mathbf{p}_{ini})}{\partial \hat{x}_{ini}} & \cdots & \frac{\partial \tilde{\mathbf{a}}_{ij}^T(\mathbf{p}_{ini})}{\partial \hat{x}_{ini}} & \cdots & \frac{\partial \tilde{\mathbf{a}}_{(L-1)L}^T(\mathbf{p}_{ini})}{\partial \hat{x}_{ini}} \end{bmatrix}^T. \tag{30}$$

$$\frac{\partial \tilde{\mathbf{a}}_{ij}(\mathbf{p}_{ini})}{\partial \hat{x}_{ini}} = \frac{\partial \Delta\tau_{ij11}}{\partial \hat{x}_{ini}} \begin{bmatrix} -j2\pi f_0 e^{-j2\pi f_0 \Delta\tau_{ij11}} \\ \vdots \\ -j2\pi f_{D(N_f-1)} e^{-j2\pi f_{D(N_f-1)} \Delta\tau_{ij11}} \end{bmatrix}. \tag{31}$$

$$\frac{\partial \Delta\tau_{ij11}}{\partial \hat{x}_{ini}} = \frac{([\mathbf{p}_{ini}]_1 - [\mathbf{S}_i]_1)}{\|\mathbf{S}_i - \mathbf{p}_{ini}\| c} - \frac{([\mathbf{p}_{ini}]_1 - [\mathbf{S}_j]_1)}{\|\mathbf{S}_j - \mathbf{p}_{ini}\| c}. \tag{32}$$

Since the partial derivative of $\boldsymbol{b}(\mathbf{p}_{ini})$ with respect to $\hat{y}_{ini}$ is of the same form as that with respect to $\hat{x}_{ini}$, no further elaboration is provided here. The final estimation of the emitter $\hat{\mathbf{p}} = [\hat{x}, \hat{y}]^T$ can be denoted as

$$\begin{bmatrix} \hat{x} \\ \hat{y} \end{bmatrix} = \begin{bmatrix} \hat{x}_{ini} \\ \hat{y}_{ini} \end{bmatrix} + \begin{bmatrix} \Delta x \\ \Delta y \end{bmatrix}. \tag{33}$$

The essential steps of the proposed algorithm are summarized below:

1. Select the reference signal $x_1(n)$ based on the prior information and segment the signals received by each sensor.
2. According to Equation (4), compute the cross-correlation between any two sensors and obtain the cross-spectrum by applying the DFT.
3. Select the effective portion of the cross-spectrum according to Equation (17), and then perform forward spatial smoothing to obtain $\tilde{\boldsymbol{R}}_{ij}$ according to Equation (18).
4. Obtain the initial position estimation $\mathbf{p}_{ini}$ based on Equations (22)–(24).
5. Obtain the final position estimation $\hat{\mathbf{p}}$ based on Equations (25) and (29)–(33).

5. Performance Analysis

5.1. Complexity Analysis

In this subsection, we contrast the computational complexity of the JTAC algorithm with that of the DPD-MFF [21] algorithm, the DPD-CS [37] algorithm, and the initial estimation of the proposed algorithm (DPD-AC). In this paper, for convenience of comparison, computational complexity is quantified by the number of complex multiplications, which is approximately four times that required for real multiplications. Define N_x and N_y as the number of grid points for the x-axis and y-axis, respectively. The computational complexity of the cross-correlation is approximately $O(N)$, and that of the cross-spectrum is $O(NlogN)$). The computational complexity for computing the covariance matrix is $O(N_f^2 D)$, and that for eigenvalue decomposition is $O(N_f^3)$. Subsequently, the computational complexity for computing the cost function is $O(L-1)LN_xN_yN_f(2N_f-1)/2$, and that of the Taylor compensation is $O(2L(L-1)(4N_f^2-N_f)+16N_f)$. Table 1 summarizes the computational complexities of these four algorithms.

Table 1. Complexity of four algorithms.

Algorithm	Computational Complexity
Proposed	$O(KLNlog_2N+K(L-1)L(N_f^2D+N)/2+(L-1)L(N_f^3+N_xN_yN_f(2N_f-1))/2+2L(L-1)(4N_f^2-N_f)+16N_f)$
DPD-AC	$O(KLNlog_2N+K(L-1)L(N_f^2D+N)/2+(L-1)L(N_f^3+N_xN_yN_f(2N_f-1))/2)$
DPD-CS	$O(KLNlog_2N+K(L-1)(N_f^2D+N)/2+(L-1)(N_f^3+N_xN_yN_f(2N_f-2M+1)))$
DPD-MFF	$O(LKNlog_2N+KN_fL^2+N_fL^3+N_xN_y(2N_f(L-M)L^2+L^3))$

Figure 3 illustrates the complexities of the four algorithms for different N_f, and the simulation conditions are $K=50$, $L=3$, $N=256$, $D=2$, $N_x=N_y=100$, $M=3$ and N_f is changing from 10 to 50. It can be seen that since both the proposed algorithm and the DPD-AC algorithm consider the complete cross-spectrum information, their computational complexity is slightly higher than that of the CS algorithm. However, this increase is minimal while yielding a significant performance improvement. Meanwhile, the proposed algorithm exhibits computational complexity nearly equivalent to that of DPD-AC, while its localization performance is greatly enhanced. On the other hand, although the DPD-MFF algorithm has the lowest computational complexity among the four, its localization performance in multipath environments is extremely poor, as will be clearly demonstrated in the subsequent simulations.

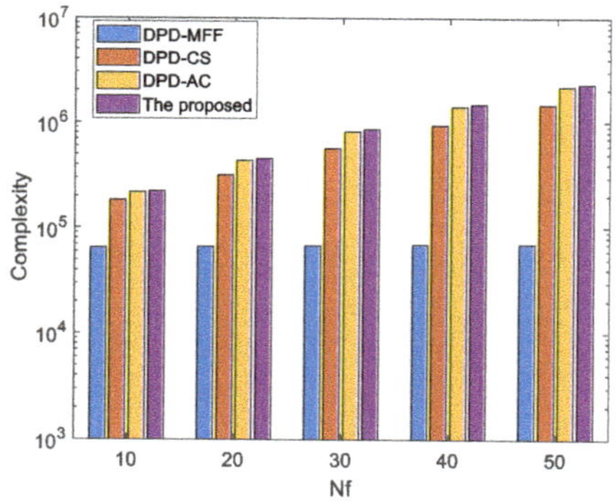

Figure 3. Computational complexity of different algorithm for the different N_f.

5.2. Effectiveness of the Proposed Algorithm

In this subsection, we use four sensors to localize an emitter to validate the effectiveness of the proposed method. In this simulation, the quadrature amplitude modulation (QAM) signal with a bandwidth of B = 32 MHz is used to simulate the emitted signal of the emitter. The simulation parameters are set as N_a = 102,400, $N=2048$, $K=50$, $N_f=40$, $M=3$, $D=10$, the sampling frequency is f_s = 125 MHz and $SNR=0$ dB. The real position of the emitter is $\mathbf{p}=[217.74\text{ m}, 340.63\text{ m}]^T$, the four sensors are located at $\mathbf{S}_1=[0\text{ m}, 0\text{ m}]^T$, $\mathbf{S}_2=[0\text{ m}, 1000\text{ m}]^T$, $\mathbf{S}_3=[1000\text{ m}, 0\text{ m}]^T$, and $\mathbf{S}_4=[1000\text{ m}, 1000\text{ m}]^T$, respectively. The simulation result with 50 repeated experiments is shown in Figure 4. It is evident that even in the low SNR, the position of the emitter can be determined by the proposed DPD-JTAC algorithm accurately.

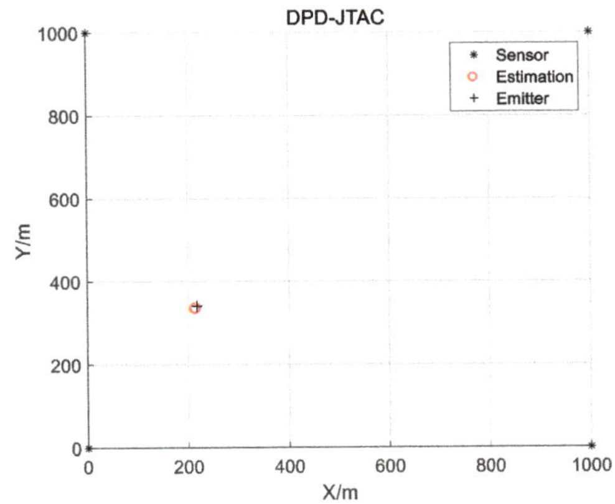

Figure 4. The simulation result.

5.3. Advantage of the Proposed Algorithm

In this subsection, we apply Monte Carlo experiments to estimate the algorithm's performance to illustrate the advantage of our proposed algorithm. And we use the root mean square error (RMSE) to evaluate the algorithm's localization accuracy, and it can be denoted as

$$\text{RMSE} = \sqrt{\frac{1}{N_m}\sum_{n_m=1}^{N_m} ||\hat{\mathbf{p}}_{n_m} - \mathbf{p}||^2}, \tag{34}$$

where $\hat{\mathbf{p}}_{n_m}$ denotes the position of emitter within the n_m-th Monte Carlo experiment. In the simulation analysis in this subsection, N_m is set to 800.

In the first simulation, we compare the performance of the proposed algorithm with the DPD-CS algorithm, the DPD-MFF algorithm, the initial estimation of the proposed algorithm (DPD-AC) and the Two-Step algorithm, under different SNRs by calculating the RMSE using Monte Carlo simulation. The parameters are set the same as Section 5.2 except for the SNR changing from −12 dB to 8 dB and the comparison result is shown in Figure 5. Observably, SNR exerts a significant influence on the proposed algorithm's performance. The performance of the proposed algorithm is improved and significantly outperforms the other three DPD algorithms as the increases. There are two basic reasons behind this. Firstly, the proposed algorithm employs forward spatial smoothing to eliminate the effects of multipath, thereby achieving performance superior to that of the unprocessed DPD-MFF algorithm. Secondly, the proposed algorithm utilizes the complete cross-spectrum information and applies Taylor compensation on the basis of a coarse estimate, resulting in performance that is undoubtedly superior to that of the DPD-CS algorithm, which uses only $L-1$ groups of cross-spectrum information, as well as the coarse-estimate DPD-AC algorithm. In addition, the other four algorithms lead to more accurate tracking results compared to the Two-Step algorithm due to the improved likelihood function using the super-resolution cost function.

In the second simulation, we analyze the impact of the arrival paths number on the localization accuracy of five algorithms by calculating the RMSE. The other parameters remain the same as the previous simulation, except the number of arrival paths, and SNR is set to 10 dB. The simulation result is shown in Figure 6. The number of arrival paths is increased from 1 to 5. It is clear that as the number of arrival paths increases, the performance of the DPD-MFF and Two-Step algorithms deteriorates drastically, whereas the performance of the other three algorithms remains nearly unchanged. Moreover, the proposed algorithm consistently maintains optimal localization performance, which indicates that it exhibits greater robustness in multipath environment.

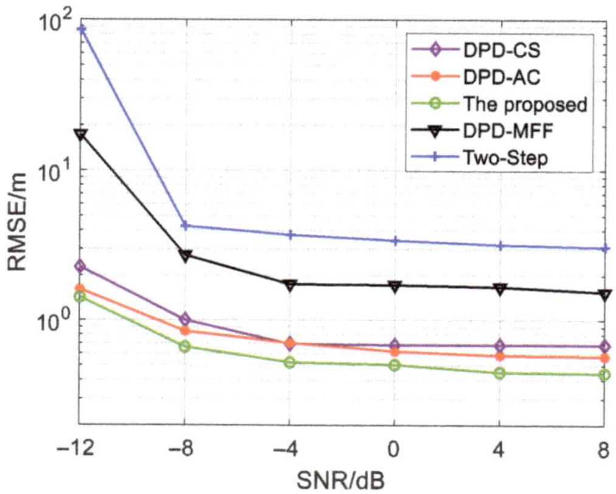

Figure 5. RMSE of various algorithms under different SNRs.

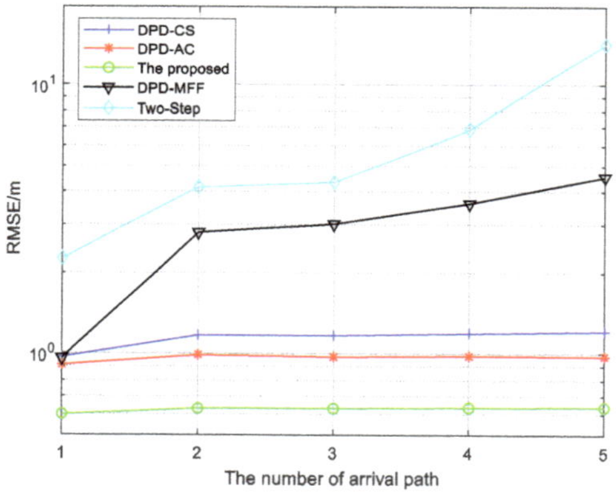

Figure 6. RMSE of various algorithms versus the number of arrival path.

Through a comprehensive comparison of Figures 5 and 6, it can be observed that the proposed algorithm effectively adapts to multipath environments, and its performance under such conditions is significantly superior to that of the other algorithms.

6. Experiment Result

In this subsection, the real-world experiment is conducted using the sensors on the campus to verify the viability of the proposed algorithm. Figure 7 is the experiment scenario. The source used in this experiment is 1465L-V Signal Generator. The sensor nodes used in the experiment are placed on the rooftops and ground of the campus, consisting of an antenna and a box equipped with node devices. Figure 8 is the schematic diagram of the node and its internal environment.

Figure 7. Source and sensing sensor distribution diagram.

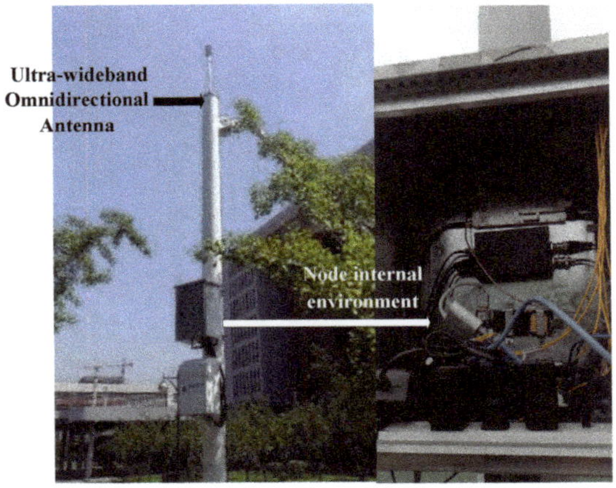

Figure 8. Sensor node and its internal environment.

To verify the effectiveness of the algorithm, we conducted 15 repeated experiments. During each experiment, the source transmitted a 32QAM signal with a bandwidth of 20 MHz and a center frequency of 720 MHz. The sampling frequency used in the experiment is $f_s = 125$ MHz, the sampling center frequency is $f_{sc} = 700$ MHz, and the sampling points during every experiment is 32,508. To facilitate the processing of the received signals, a Cartesian coordinate system is established based on the latitudes and longitudes of the four sensors and the emitter's location, with Sensor 2 serving as the coordinate origin. The coordinates of four sensors are $\mathbf{S}_1 = [-60.1265, 32.06]^T$, $\mathbf{S}_2 = [0,0]^T$, $\mathbf{S}_3 = [23.5975, -95.4008]^T$, $\mathbf{S}_4 = [5.3802, -188.353]^T$.

Figure 9 shows the frequency spectrum of the signals received by the four sensors in one experiment, from which the wideband signal can be clearly seen. The localization results of the source with the proposed JTAC algorithm is shown in Figure 10. Clearly, even in a complex and dynamic real electromagnetic environment, the proposed algorithm is also capable of estimating the source's position.

To highlight the advantages of the proposed algorithm, we compared the cumulative distribution function (CDF) curves of localization errors for different algorithms as shown in Figure 11. In comparison to other algorithms, the proposed method aligns most closely with the vertical axis, with almost 90% of the error values falling below 20 m, which demonstrates that the direct positioning algorithm proposed in this paper exhibits lower error compared to other algorithms, indicating its superior localization performance.

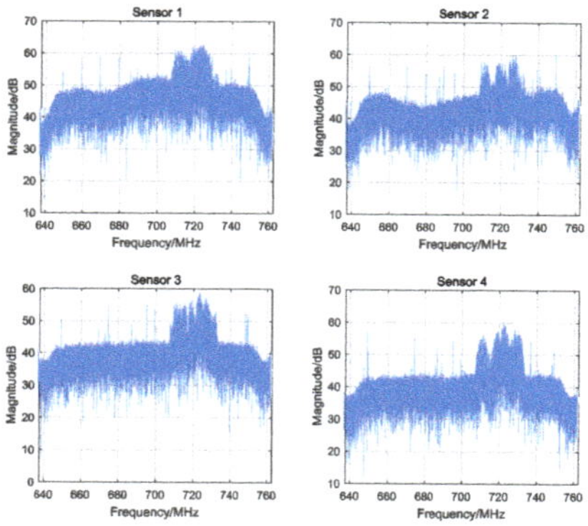

Figure 9. The frequency spectrum of the signals received by the four sensors.

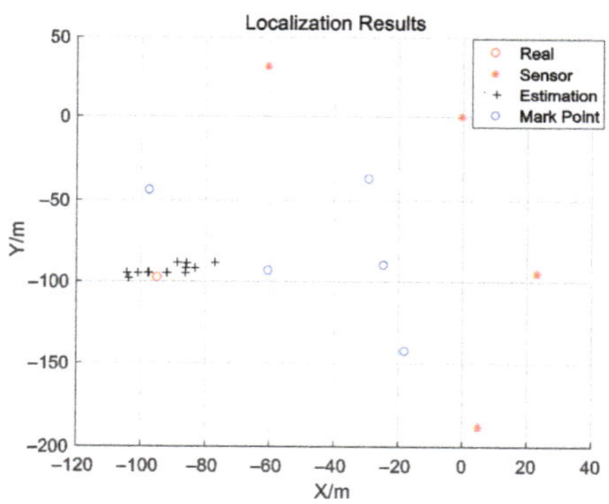

Figure 10. The localization results.

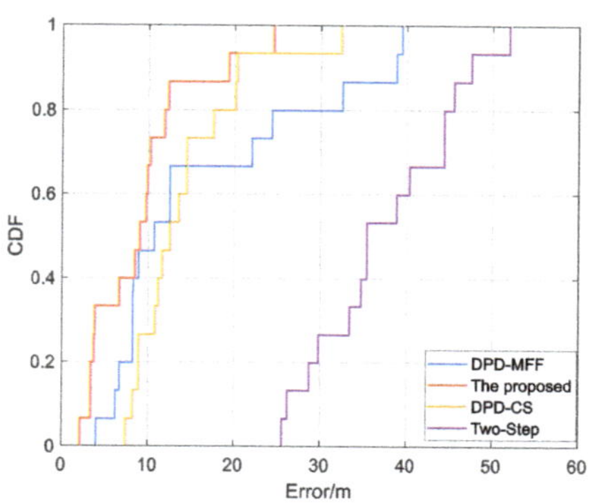

Figure 11. Error CDF curves of different algorithms.

7. Conclusions

In this paper, a direct position determination method for wideband sources over multipath environments is proposed. Our proposed method introduces a better data model based on the cross-spectrum, which utilizes the received signals from any pair of sensors. Then, we employ forward spatial smoothing to eliminate the multipath effects. Moreover, Taylor expansion is used to obtain the optimal compensation value, enhancing the localization accuracy when the source is not located on the search grid. Finally, the performance analysis, including numerical simulation and real-world experiment, demonstrates the superior localization capability of the proposed method in comparison with existing approaches.

Author Contributions: Conceptualization, X.Z.; methodology, H.C. and X.Y.; software, H.C. and X.Y.; investigation, H.H.; data curation, X.Y.; writing—original draft preparation, H.C. and X.Y.; writing—review and editing, X.Z. All authors have read and agreed to the published version of the manuscript.

Funding: This research was supported by China NSF Grants (62371225), China NSF Grants (62371227) and China NSF Grants (62427801).

Institutional Review Board Statement: Not applicable.

Data Availability Statement: The datasets presented in this article are not readily available because the data are part of an ongoing study. Requests to access the datasets should be directed to xinjianyin@nuaa.edu.cn.

Conflicts of Interest: The authors declare no conflicts of interest.

Abbreviations

The following abbreviations are used in this manuscript:

DPD	Direct Position Determination
DOA	Directory of Arrival
TOA	Time of Arrival
TDOA	Time Difference in Arrival
RSS	Received Signal Strength
FDOA	Frequency Difference in Arrival
LS	Least Square
ML	Maximum Likelihood
SDF	Subspace Data Fusion
MUSIC	Multiple Signal Classification
MVDP	Minimum Variance Distortionless Response
QPSO	Quantum-behaved Particle Swarm Optimization
MDF	Multi-array Data Fusion
SNR	Signal-to-Noise Ratio
OWLS	Optimal Weighted Lest Square
FIM	Fisher Information Matrix
CRB	Cramér–Rao Bound
LFM	Linear Frequency Modulation
STFT	Short-Time Fourier Transform
HT	Hough Transform
MFF	Multiple-frequency Function Fusion
PARAFAC	Parallel Factor

References

1. Subedi, S.; Zhang, Y.D.; Amin, M.G.; Himed, B. Group sparsity based multi-target tracking in passive multi-static radar systems using Doppler-only measurements. *IEEE Trans. Signal Process.* **2016**, *64*, 3619–3634. [CrossRef]
2. Tang, Z.; Manikas, A. Multi direction-of-arrival tracking using rigid and flexible antenna arrays. *IEEE Trans. Wirel. Commun.* **2021**, *20*, 7568–7580. [CrossRef]
3. Chervoniak, Y.; Sinitsyn, R.; Yanovsky, F.; Zaporozhets, O. TDOA and Doppler Shift Estimation Method for Passive Acoustic Location of Flying Vehicles. In Proceedings of the 2018 IEEE 17th International Conference on Mathematical Methods in Electromagnetic Theory (MMET), Kyiv, Ukraine, 2–5 July 2018; IEEE: New York, NY, USA, 2018; pp. 119–122.
4. Ye, H.; Yang, B.; Long, Z.; Dai, C. A method of indoor positioning by signal fitting and PDDA algorithm using BLE AOA device. *IEEE Sens. J.* **2022**, *22*, 7877–7887. [CrossRef]
5. Shen, J.; Molisch, A.F.; Salmi, J. Accurate passive location estimation using TOA measurements. *IEEE Trans. Wirel. Commun.* **2012**, *11*, 2182–2192. [CrossRef]
6. Li, Q.; Chen, B.; Yang, M. Improved two-step constrained total least-squares TDOA localization algorithm based on the alternating direction method of multipliers. *IEEE Sens. J.* **2020**, *20*, 13666–13673. [CrossRef]
7. Kim, D.G.; Park, G.H.; Kim, H.N.; Park, J.O.; Park, Y.M.; Shin, W.H. Computationally efficient TDOA/FDOA estimation for unknown communication signals in electronic warfare systems. *IEEE Trans. Aerosp. Electron. Syst.* **2017**, *54*, 77–89. [CrossRef]
8. Li, X. RSS-based location estimation with unknown pathloss model. *IEEE Trans. Wirel. Commun.* **2006**, *5*, 3626–3633. [CrossRef]
9. You, M.Y.; Lu, A.N.; Ye, Y.X.; Huang, K.; Lou, C. Direct Position Determination Using Compressive Sensing Measurements without Reconstruction. *IEEE Trans. Aerosp. Electron. Syst.* **2022**, *59*, 2036–2043. [CrossRef]
10. Weiss, A.J. Direct position determination of narrowband radio frequency transmitters. *IEEE Signal Process. Lett.* **2004**, *11*, 513–516. [CrossRef]
11. Demissie, B.; Oispuu, M.; Ruthotto, E. Localization of multiple sources with a moving array using subspace data fusion. In Proceedings of the 2008 11th International Conference on Information Fusion, Cologne, Germany, 30 June–3 July 2008; IEEE: New York, NY, USA, 2008; pp. 1–7.
12. Schmidt, R. Multiple emitter location and signal parameter estimation. *IEEE Trans. Antennas Propag.* **1986**, *34*, 276–280. [CrossRef]
13. Tzafri, L.; Weiss, A.J. High-resolution direct position determination using MVDR. *IEEE Trans. Wirel. Commun.* **2016**, *15*, 6449–6461. [CrossRef]
14. Yuanchun Tang, R.C.; Xia, B. Direct positioning method of interference source of satellite navigation system based on propagation operator. *J. Terahertz Sci. Electron. Inf. Technol.* **2023**, *21*, 985–991.
15. Huang, Z.; Wu, J. Multi-array data fusion based direct position determination algorithm. In Proceedings of the 2014 Seventh International Symposium on Computational Intelligence and Design, Hangzhou, China, 13–14 December 2014; IEEE: New York, NY, USA, 2014; Volume 2, pp. 121–124.
16. Sun, Y.; Zhang, X.; Shi, X. Direct Position Determination Algorithm for Multiple Arrays via Weighted Propagator Method. *China Commun.* **2022**, 1–13. [CrossRef]
17. Shi, X.; Zhang, X.; Qian, Y.; Ahmad, M. Closed-form rectilinear emitters localization with multiple sensor arrays: Phase alignment and optimal weighted least-square method. *IEEE Sens. J.* **2023**, *23*, 7266–7278. [CrossRef]
18. Vankayalapati, N.; Kay, S.; Ding, Q. TDOA based direct positioning maximum likelihood estimator and the Cramer-Rao bound. *IEEE Trans. Aerosp. Electron. Syst.* **2014**, *50*, 1616–1635. [CrossRef]
19. Chen, F.; Zhou, T.; Yi, W.; Kong, L.; Zhai, B. Passive direct position determination of multiple emitters transmitting unknown LFM signals. In Proceedings of the 2018 IEEE Radar Conference (RadarConf18), Oklahoma City, OK, USA, 23–27 April 2018; IEEE: New York, NY, USA, 2018; pp. 1129–1133.
20. Hao, K.; Wan, Q. An efficiency-improved TDOA-based direct position determination method for multiple sources. In Proceedings of the ICASSP 2019—2019 IEEE International Conference on Acoustics, Speech and Signal Processing (ICASSP), Brighton, UK, 12–17 May 2019; IEEE: New York, NY, USA, 2019; pp. 4425–4429.
21. Zhu, K.; Jiang, H.; Yu, L.; Li, Y.; Li, J. An Improved TDOA-based DPD Method via Multiple-frequency Function Fusion. In Proceedings of the 2022 7th International Conference on Signal and Image Processing (ICSIP), Suzhou, China, 20–22 July 2022; IEEE: New York, NY, USA, 2022; pp. 217–221.
22. Li, J.; Li, Y.; Jiang, H.; Zhang, X.; Wu, Q. Multi-TDOA estimation and source direct position determination based on parallel factor analysis. *IEEE Internet Things J.* **2022**, *10*, 7405–7415. [CrossRef]
23. Li, D.; Li, J.; You, M.; Tang, W.; Zhang, X.; Qin, W. Direct localization of wideband sources using distributed arrays: A subspace focusing and dimension reduction approach. *Signal Process.* **2025**, *232*, 109902. [CrossRef]
24. Piccinni, G.; Avitabile, G.; Coviello, G.; Talarico, C. Real-time distance evaluation system for wireless localization. *IEEE Trans. Circuits Syst. I Regul. Pap.* **2020**, *67*, 3320–3330. [CrossRef]

25. Li, S.; Qi, C.; Zhang, Y. A low computation direct position determination based on TDOA. In Proceedings of the 2021 3rd International Academic Exchange Conference on Science and Technology Innovation (IAECST), Guangzhou, China, 10–12 December 2021; IEEE: New York, NY, USA, 2021; pp. 63–67.
26. Qi, S.R.; Yuan, C.S.; Liang, B.J.; Lu, G.; Hu, L.; Heng, S. A direct position determination method for four-station TDOA location system. In Proceedings of the 2019 IEEE International Conference on Signal, Information and Data Processing (ICSIDP), Chongqing, China, 11–13 December 2019; IEEE: New York, NY, USA, 2019; pp. 1–5.
27. Ma, F.; Liu, Z.M.; Guo, F. Direct position determination in asynchronous sensor networks. *IEEE Trans. Veh. Technol.* **2019**, *68*, 8790–8803. [CrossRef]
28. Zhong, S.; Xia, W.; Song, J.; He, Z. Super-resolution time delay estimation in multipath environments using normalized cross spectrum. In Proceedings of the 2013 International Conference on Communications, Circuits and Systems (ICCCAS), Chengdu, China, 15–17 November 2013; IEEE: New York, NY, USA, 2013; Volume 1, pp. 288–291.
29. Ge, F.X.; Shen, D.; Peng, Y.; Li, V.O. Super-resolution time delay estimation in multipath environments. *IEEE Trans. Circuits Syst. I Regul. Pap.* **2007**, *54*, 1977–1986. [CrossRef]
30. Li, J.; Li, P.; Li, P.; Tang, L.; Zhang, X.; Wu, Q. Self-position awareness based on cascade direct localization over multiple source data. *IEEE Trans. Intell. Transp. Syst.* **2022**, *25*, 796–804. [CrossRef]
31. Liu, Y.; Shi, X. DPD of NC signals with multiple base stations: Reduced Dimension Propagator Method and Taylor compensation. *J. Terahertz Sci. Electron. Inf. Technol.* **2023**, *21*, 725–733.
32. Chen, K.; Chen, W.; Li, J. Noncircular Distributed Source DOA Estimation with Nested Arrays via Reduced-Dimension MUSIC. *Sensors* **2024**, *24*, 6653. [CrossRef] [PubMed]
33. Gao, Y.; Zhang, X.; You, M.; Pan, H.; Mao, S.; Jiang, K.; Li, J. Direct Self-Position Determination Exploiting Multiple Non-circular Emitters: A Reduced Dimensional Weighted Propagator Method. *IEEE Sens. J.* **2024**, *24*, 27836–27846. [CrossRef]
34. Zhu, K.; Jiang, H.; Li, J.; Zhou, F. A direct position determination method using distributed uavs with synchronization error. *IEEE Sens. J.* **2023**, *24*, 780–787. [CrossRef]
35. Knapp, C.; Carter, G. The generalized correlation method for estimation of time delay. *IEEE Trans. Acoust. Speech Signal Process.* **1976**, *24*, 320–327. [CrossRef]
36. Li, J.; Zhu, K.; Jiang, H.; Zhang, X.; Wu, Q. A Direct Position Determination Method of TDOA Based on Multi-Frequency Focusing and Fusion for Coherent Radiation Sources. *Acta Electron. Sin.* **2023**, *51*, 2110–2117.
37. Zhu, K.; Jiang, H.; Huo, Y.; Yu, Q.; Li, J. A direct position determination method based on subspace orthogonality in cross-spectra under multipath environments. *Sensors* **2022**, *22*, 7245. [CrossRef]

sensors

MDPI

Article

Active RIS-Assisted Uplink NOMA with MADDPG for Remote State Estimation in Wireless Sensor Networks

Rongzhen Li † and Lei Xu *,†

School of Computer Science and Engineering, Nanjing University of Science and Technology, Nanjing 210094, China; lirongzhen0213@163.com
* Correspondence: xulei_marcus@126.com
† These authors contributed equally to this work.

Abstract: Non-orthogonal multiple access (NOMA) and reconfigurable intelligent surfaces (RISs) are recognized as key technologies for beyond 5G and 6G wireless communications. To address the high computational complexity and non-convex optimization challenges, this letter proposes an optimization framework based on the Multi-Agent Deep Deterministic Policy Gradient (MADDPG) algorithm. The proposed framework jointly makes use of sensor grouping, power allocation, an RIS computation strategy, and phase shifts to minimize the remote state estimation (RSE) error. Simulation results demonstrate that the MADDPG algorithm, when applied in an RIS-assisted NOMA system, significantly reduces the RSE error.

Keywords: wireless sensor networks; uplink NOMA; RIS; MADDPG algorithm; RSE error

1. Introduction

With the increasing number of wireless devices and the development of ultra-5G and 6G technologies in industrial Internet of Things (IoT) networks, wireless sensor networks (WSNs) [1], as a key component of IoT systems, face higher demands for low latency, low remote state estimation (RSE) errors, and low system error rates (SERs) [2–4]. As the number of smart devices grows and large-scale IoT applications proliferate, ensuring the efficiency and reliability of these systems has become a critical issue to address.

To enhance spectrum efficiency in WSNs, non-orthogonal multiple access (NOMA) technology has been widely adopted due to its capability to support multiple sensors with the same time–frequency resources through superposition coding and successive interference cancellation [5,6]. Recent studies have enhanced NOMA further by integrating active reconfigurable intelligent surfaces (RISs) to boost the link reliability and energy efficiency. For instance, the work presented in [7] investigates joint UAV trajectory optimization and active RIS control in NOMA systems, highlighting significant energy efficiency gains. However, this study primarily addressed aerial–terrestrial network scenarios and was not directly applicable to ground-based sensor networks.

Both passive and active RIS technologies have been explored extensively for improving NOMA system performance. Recent research, such as that presented in [8], demonstrates passive RISs' effectiveness in reducing the bit error rate (BER) and extending the network coverage. Nevertheless, a passive RIS lacks active amplification capabilities and adaptability under highly dynamic channel conditions. Conversely, an active RIS, as explored in [9], enhances the physical-layer security by providing analytical derivations for the secrecy outage probability. However, this study does not incorporate critical aspects such as sensor grouping strategies or estimation error optimization.

Traditional optimization methods like block coordinate descent (BCD) and successive convex approximation (SCA) have been employed for resource allocation in RIS-assisted NOMA networks. For example, the research by Xu et al. [10] achieves improved system sum rates using BCD-based resource scheduling techniques. Additionally, recent efforts such as [11] have explored low-complexity signal processing schemes, including sparse channel parameter estimation for MIMO-FBMC systems, which are highly relevant to industrial big data communication scenarios. However, these approaches typically focus on the physical layer and do not address joint decision-making strategies or cross-layer coordination for sensor estimation. Moreover, they lack multi-agent coordination mechanisms and considerations for remote state estimation (RSE) error control, which are essential in practical WSN deployments.

Meanwhile, reinforcement learning methodologies, including deep reinforcement learning (DRL), deep deterministic policy gradient (DDPG), and multi-agent DDPG (MADDPG), have emerged as effective solutions for dynamic resource allocation. A recent IEEE Transactions study [12] presents an attention-augmented MADDPG framework applied to NOMA-based vehicular mobile edge computing, demonstrating improved convergence rates and predictive accuracy under realistic workload scenarios. Additionally, recent studies such as [13] have employed DRL approaches to jointly optimize the UAV trajectories, transmission power, and RIS phase shifts in RIS-aided UAV-NOMA systems. Nonetheless, these studies have predominantly considered passive RIS configurations and assumed static user grouping, limiting the adaptability in dynamic scenarios.

To the best of our knowledge, no existing research has jointly tackled the sensor state estimation error (RSE), sensor grouping, transmission power control, and active RIS beamforming within an uplink NOMA-assisted WSN using a decentralized multi-agent reinforcement learning framework. To this end, we propose a MADDPG-based distributed policy learning approach. MADDPG is particularly suitable for this problem setting because it explicitly models the interactions among agents and handles decentralized control in partially observable environments. In our scenario, each sensor acts as an agent, making decisions based only on its local state, such as channel gain and estimation covariance. This naturally forms a Dec-POMDP (Decentralized Partially Observable Markov Decision Process), to which MADDPG is well suited. It enables coordinated learning during centralized training while maintaining independent execution, supports hybrid discrete–continuous action spaces, and offers scalability in large sensor networks with dynamic channel conditions.

The main contributions of this paper are summarized as follows:

1. Joint Integration of an Active RIS and MADDPG: We propose a joint optimization framework that integrates an active RIS and MADDPG-based decentralized learning for uplink WSNs. This design improves the estimation accuracy, reduces the power consumption, and enhances the system's adaptability.
2. Real-Time Estimation via Kalman Filtering: The proposed method incorporates Kalman filtering (KF) to enable real-time state tracking, which improves the estimation robustness under varying network conditions.
3. Decentralized Resource Allocation: We formulate a novel decentralized resource allocation problem that jointly considers sensor grouping, power control, and active RIS beamforming to minimize the RSE error—an optimization objective rarely addressed in existing works. Each sensor acts as an independent agent and makes decisions based on its local state in a Dec-POMDP setting. The MADDPG framework is employed to enable scalable and coordinated learning among agents. While our method is built upon an established RL architecture, it is tailored to a unique estimation-driven

optimization task in WSNs. We also acknowledge the absence of a formal theoretical convergence analysis and highlight this as a direction for future investigation.

4. Performance Gains Verified by Simulations: Simulation results show significant improvements in the RSE error reduction and system reliability compared to those under conventional and single-agent approaches, confirming the practicality of the proposed framework.

The remainder of this paper is organized as follows: Section 2 details the local sensor state estimation and the uplink NOMA transmission model with an active RIS, including the derivation of the RSE error metrics and the outage probability model. Section 3 introduces the MADDPG-based decentralized algorithm, elaborating on the joint optimization framework that integrates sensor grouping, power allocation, and active RIS beamforming, as well as incorporating KF or enhanced real-time state estimation. Section 4 presents the simulation experiments, performance comparisons with benchmark algorithms, and a comprehensive analysis of the results demonstrating improvements in the RSE accuracy and network reliability. Finally, Section 5 concludes this paper by summarizing the key findings and potential directions for future research.

Notations: $\mathbb{R}$ denotes the set of real numbers. $\mathbb{C}$ represents the set of complex numbers. The symbol $(\cdot)^{\mathrm{T}}$ denotes the transpose operation, and the superscript $(\cdot)^{-1}$ indicates the matrix inverse. The notation $\mathbb{P}[\cdot]$ refers to probability and $\mathbb{E}[\cdot]$ to expectation. $\mathrm{Tr}(\cdot)$ represents the trace operator of a matrix. A bold symbol (e.g., $\mathbf{P}$) is used to denote vectors or matrices, with lowercase bold letters representing vectors and uppercase bold letters representing matrices. Non-bold symbols (e.g., p) indicate scalar quantities such as real-valued parameters or constants.

2. The System Model

This letter explores an RSE system within an uplink NOMA transmission model, aided by an active RIS. As depicted in Figure 1, the system consists of a single BS, equipped with G receive antennas, an active RIS with M reflective elements and N single-antenna sensors. The sets of sensors, sensor groups, and reflective elements are denoted as $\mathcal{N} = \{1, 2, \cdots, N\}$, $\mathcal{G} = \{1, 2, \cdots, G\}$ and $\mathcal{M} = \{1, 2, \cdots, M\}$, respectively.

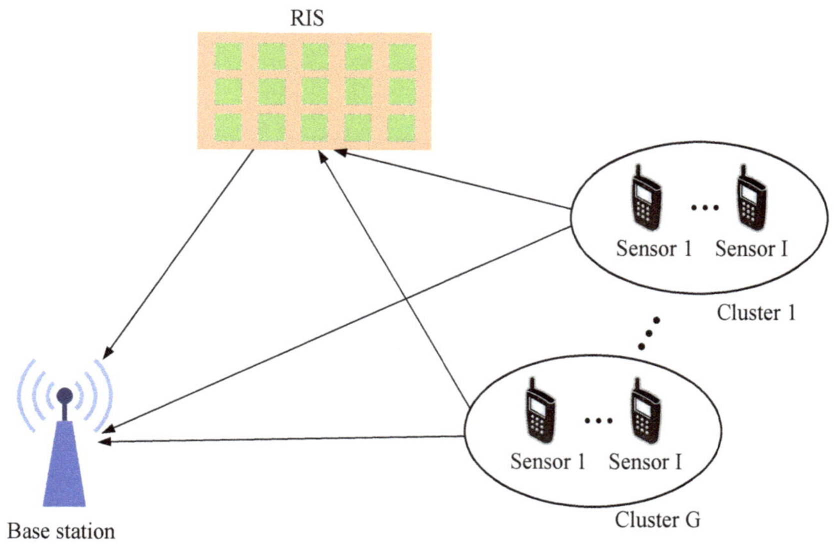

Figure 1. Active RIS-assisted uplink NOMA model.

This letter explores an RSE system within an uplink NOMA transmission model, aided by an active RIS. As depicted in Figure 1, the system consists of a single BS equipped with G receive antennas; an active RIS with M reflective elements; and N single-antenna sensors. To facilitate decentralized decision-making, the N sensors are partitioned into G sensor groups, each served by the BS equipped with G antennas. Note that the grouping does not imply a one-to-one mapping between each group and a specific antenna; instead, it is designed for efficient multi-agent training and resource allocation. The sets of sensors, sensor groups, and reflective elements are denoted as $\mathcal{N} = \{1, 2, \cdots, N\}$, $\mathcal{G} = \{1, 2, \cdots, G\}$, and $\mathcal{M} = \{1, 2, \cdots, M\}$, respectively.

2.1. Local State Estimation

The output observed by sensor n at time k for the associated linear time-invariant (LTI) process can be expressed as follows

$$\mathbf{x}_n^{k+1} = \mathbf{A}_n \mathbf{x}_n^k + \mathbf{w}_n^k, \tag{1}$$

$$\mathbf{y}_n^{k+1} = \mathbf{C}_n \mathbf{x}_n^k + \mathbf{v}_n^k, \tag{2}$$

where $\mathbf{x}_n^k \in \mathbb{R}^{l_n}$ and $\mathbf{y}_n^k \in \mathbb{R}^{r_n}$ represent the state and measurement vectors, respectively [14], with l_n and r_n denoting the dimensions of the state and measurement vectors for sensor n. The state evolution is governed by the state transition matrix $\mathbf{A}_n \in \mathbb{R}^{l_n \times l_n}$, while the measurement process is characterized by the measurement matrix $\mathbf{C}_n \in \mathbb{R}^{r_n \times l_n}$. The disturbances $\mathbf{w}_n^k \in \mathbb{R}^{l_n}$ and $\mathbf{v}_n^k \in \mathbb{R}^{r_n}$ are assumed to be zero-mean, independent, and identically distributed (i.i.d.) Gaussian noise processes, with covariance matrices $\mathbf{Q}_n$ and $\mathbf{R}_n$, respectively.

Each sensor operates a local Kalman filter (KF) and transmits the corresponding minimum mean squared error (MMSE) state estimate to the remote estimator, along with the associated remote state estimation (RSE) error covariance $\mathbf{P}_n^{k|k}$, which are defined as [15]

$$\hat{\mathbf{x}}_n^{k|k} \triangleq \mathbb{E}\left[\mathbf{x}_n^k \mid \mathbf{y}_n^1, \mathbf{y}_n^2, \ldots, \mathbf{y}_n^k\right], \tag{3}$$

$$\mathbf{P}_n^{k|k} \triangleq \mathbb{E}\left[\left(\mathbf{x}_n^k - \hat{\mathbf{x}}_n^{k|k}\right)\left(\mathbf{x}_n^k - \hat{\mathbf{x}}_n^{k|k}\right)^\top\right]. \tag{4}$$

The prior state estimate $\hat{\mathbf{x}}_n^{k|k-1}$ and the posterior state estimate $\hat{\mathbf{x}}_n^{k|k}$ at time k are updated through the following steps. The Kalman gain is denoted by $\mathbf{K}_n^k$, and the error covariances are represented by $\mathbf{P}_n^{k|k-1}$ and $\mathbf{P}_n^{k|k}$, corresponding to the prior and posterior error covariances, respectively. The KF update process is given by

$$\hat{\mathbf{x}}_n^{k|k-1} = \mathbf{A}_n \hat{\mathbf{x}}_n^{k-1|k-1}, \tag{5}$$

$$\mathbf{P}_n^{k|k-1} = \mathbf{A}_n \mathbf{P}_n^{k-1|k-1} \mathbf{A}_n^\top + \mathbf{Q}_n, \tag{6}$$

$$\mathbf{K}_n^k = \mathbf{P}_n^{k|k-1} \mathbf{C}_n^\top \left(\mathbf{C}_n \mathbf{P}_n^{k|k-1} \mathbf{C}_n^\top + \mathbf{R}_n\right)^{-1}, \tag{7}$$

$$\hat{\mathbf{x}}_n^{k|k} = \hat{\mathbf{x}}_n^{k|k-1} + \mathbf{K}_n^k \left(\mathbf{y}_n^k - \mathbf{C}_n \hat{\mathbf{x}}_n^{k|k-1}\right), \tag{8}$$

$$\mathbf{P}_n^{k|k} = \left(\mathbf{I}_{l_n} - \mathbf{K}_n^k \mathbf{C}_n\right) \mathbf{P}_n^{k|k-1}, \tag{9}$$

where $\mathbf{I}_{l_n}$ is the $l_n \times l_n$ identity matrix, and l_n denotes the dimension of the state vector $\mathbf{x}_n^k$.

The local estimation error covariance converges exponentially to a steady-state value. Without loss of generality, we assume that each sensor's KF reaches this steady state. To simplify the following analysis, we denote the steady-state covariance as $\mathbf{P}_n^k = \bar{\mathbf{P}}_n, k \geqslant 1$.

2.2. The Uplink Communication Model

After obtaining $\hat{x}_n^k$, sensor n transmits it to the BS as a data packet. The communication between the sensors and the BS is based on an uplink NOMA communication model, supported by an active RIS. In this model, the RIS serves as an enhancement tool for signal propagation, optimizing both the signal quality and coverage.

The equivalent baseband time-domain channel in the system is composed of three parts: the channel from the RIS to the BS, from sensor n to the RIS, and from sensor n to the BS. These channels are denoted by $\mathbf{h}^{rb} \in \mathbb{C}^{G\times M}$, $\mathbf{h}_n^{sr} \in \mathbb{C}^{M\times 1}$, and $\mathbf{h}_n^{sb} \in \mathbb{C}^{G\times 1}$, respectively. The reflection coefficient of the RIS element m is expressed as $\theta_m = \alpha_m e^{j\phi_m}$, where α_m represents the reflection amplitude and $\phi_m \in [0, 2\pi]$ denotes the phase shift. The RIS beamforming is then captured by the reflection coefficient matrix $\mathbf{\Theta} \triangleq \text{diag}\{\theta_1, \theta_2, \ldots, \theta_M\}$ [2]. Unlike passive RISs, active RISs are equipped with amplifiers that consume additional power, making thermal noise at the active RIS significant. Consequently, the signal received at the BS can be modeled as

$$\mathbf{y}_g^k = \sum_{n=1}^{N} b_{n,g}^k \sqrt{p_n^k}\mathbf{h}_n^k f_n^k + \mathbf{n}_g^k, \tag{10}$$

where $b_{n,g}^k \in \{0,1\}$ represents the binary channel selection for sensor n. The total channel gain $\mathbf{h}_n^k$ is denoted as $\mathbf{h}_k^{rb}\mathbf{\Theta}_k\mathbf{h}_{n,k}^{sr} + \mathbf{h}_{n,k}^{sb}$, $\mathbf{h}_n^k \in \mathbb{C}^{G\times 1}$. The noise $\mathbf{n}_g^k$ is defined as $\mathbf{h}_k^{rb}\mathbf{\Theta}_k\mathbf{v}_k + \mathbf{n}_{0,g}^k$, $\mathbf{n}_g^k \in \mathbb{C}^{G\times 1}$. The thermal noise at the RIS is modeled by $\mathbf{v}_k \in \mathbb{C}^{M\times 1}$, which follows a circularly symmetric complex Gaussian distribution, i.e., $\mathbf{v}_k \sim \mathcal{CN}(\mathbf{0}_M, \sigma_1^2\mathbf{I}_M)$. Similarly, the thermal noise at the BS is modeled by $\mathbf{n}_{0,g}^k \in \mathbb{C}^{G\times 1}$, where $\mathbf{n}_{0,g}^k \sim \mathcal{CN}(\mathbf{0}_G, \sigma_2^2\mathbf{I}_G)$ [2]. Let f_n^k represent the data symbol transmitted by sensor n at time k, while p_n^k denotes the corresponding power used for computation offloading. We assume that the transmitted symbols are independent and identically distributed (i.i.d.) as $f_n^k \sim \mathcal{CN}(0,1)$. Since it is an uplink NOMA channel, we assume that the channel gains between different sensors decrease, i.e., $|\mathbf{h}_1^k|^2 \geq |\mathbf{h}_2^k|^2 \cdots \geq |\mathbf{h}_n^k|^2$.

The channel model comprises both path loss and small-scale fading. For the direct link between the BS and the sensor n, we assume Rayleigh fading. The corresponding channel can be expressed as

$$\mathbf{h}_{n,k}^{sb} = \sqrt{d_{S,n,k}^{-\alpha_S}}\widetilde{\mathbf{h}}_{n,k}^{sb}, \tag{11}$$

where $d_{S,n,k}$ is the distance between the BS and sensor n, α_S is the path loss exponent, and $\widetilde{\mathbf{h}}_{n,k}^{sb} \in \mathbb{C}^{G\times 1}$ represents the small-scale fading channel, modeled as $\mathcal{CN}(\mathbf{0}_G, \mathbf{I}_G)$.

The active RIS is deployed at a fixed location, where line-of-sight (LoS) links exist from both the BS to the RIS and from the RIS to sensor n. Therefore, the reflection channel follows Rician fading. The channel matrix from the RIS to the BS, $\mathbf{h}_k^{rb}$, and the channel vector from sensor n to the RIS, $\mathbf{h}_{n,k}^{sr}$, are given by [16]

$$\mathbf{h}_k^{rb} = \sqrt{d_{B,k}^{-\alpha_B}}\left(\sqrt{\frac{\delta_B}{\delta_B+1}}\widetilde{\mathbf{h}}_k^{rb,l} + \sqrt{\frac{1}{\delta_B+1}}\widetilde{\mathbf{h}}_k^{rb}\right), \tag{12}$$

$$\mathbf{h}_{n,k}^{sr} = \sqrt{d_{R,n,k}^{-\alpha_R}}\left(\sqrt{\frac{\delta_R}{\delta_R+1}}\widetilde{\mathbf{h}}_{n,k}^{sr,l} + \sqrt{\frac{1}{\delta_R+1}}\widetilde{\mathbf{h}}_{n,k}^{sr}\right), \tag{13}$$

where $d_{B,k}$ and $d_{R,n,k}$ represent the distances from the BS to the RIS and from the RIS to sensor n, respectively. α_B and α_R are the path loss exponents, while δ_B and δ_R denote the Rician factors. The terms $\widetilde{\mathbf{h}}_k^{rb,l} \in \mathbb{C}^{G\times M}$ and $\widetilde{\mathbf{h}}_{n,k}^{sr,l} \in \mathbb{C}^{M\times 1}$ represent the LoS components from the RIS to the BS and from the sensor to the RIS, respectively. The small-scale fading components, $\tilde{\mathbf{h}}_k^{rb} \in \mathbb{C}^{G\times M}$ and $\tilde{\mathbf{h}}_{n,k}^{sr} \in \mathbb{C}^{M\times 1}$, follow the complex Gaussian distributions $\mathcal{CN}(\mathbf{0}_{G\times M}, \mathbf{I}_{G\times M})$ and $\mathcal{CN}(\mathbf{0}_M, \mathbf{I}_M)$, respectively.

Therefore, according to the uplink NOMA decoding protocol with successive interference cancellation (SIC), the signal from sensor n is decoded after those from the sensors with smaller indices (i.e., $j < n$), whose signals are assumed to have been successfully decoded and removed at the BS. Thus, the residual interference with sensor n originates only from the sensors with indices greater than n. The signal-to-interference-plus-noise ratio (SINR) of sensor n is given by

$$\gamma_n^k = \frac{p_n^k \left\| \mathbf{h}_n^k \right\|^2}{\sum_{j=n+1}^{N} b_{j,g}^k p_j^k \left\| \mathbf{h}_j^k \right\|^2 + \sigma_k^2}, \tag{14}$$

where $\sigma_k^2 \triangleq \sigma_{1,k}^2 \left\| \mathbf{h}_k^{rb} \boldsymbol{\Theta}_k \right\|^2 + \sigma_{2,k}^2$ accounts for the effective noise at the BS, including thermal noise from both the RIS and the BS receiver.

In digital communication theory, the SER for a single sensor is expressed in terms of the SINR as SER $= Q(\sqrt{\gamma})$, where $Q(x) = \frac{1}{\sqrt{2\pi}} \int_x^{\infty} e^{-\frac{t^2}{2}}\, dt$ denotes the Gaussian Q-function. By applying the above theory, the SER can be approximated as SER $\approx e^{-\gamma}$ [3]. For sensor n at time k, the SER is approximated as

$$\mathrm{SER}_n^k \approx e^{-\gamma_n^k}. \tag{15}$$

2.3. Remote State Eatimation

Each sensor transmits its local state estimation measurement data to the remote estimator (i.e., the BS) via a wireless channel, utilizing uplink NOMA technology enhanced by RIS assistance. Thus, the transmission of $\hat{\mathbf{x}}_n^k$ can be characterized by a binary random process $\left\{\lambda_n^k\right\}$, where (see Figure 2)

$$\lambda_n^k = \begin{cases} 1, & \text{if } \hat{\mathbf{x}}_n^k \text{ arrives error-free at time } k, \\ 0, & \text{otherwise (regarded as dropout).} \end{cases} \tag{16}$$

Combining (15) and (16), the probability of the successful transmission of $\hat{\mathbf{x}}_n^k$ is expressed as

$$\mathbb{P}\left[\lambda_n^k = 1\right] = 1 - \mathrm{SER}_n^k = 1 - e^{-\gamma_n^k}. \tag{17}$$

Then, the RSE error covariance is [3]

$$\mathbf{P}_n^k = \begin{cases} e^{\alpha} \bar{\mathbf{P}}_n, & \text{if } \lambda_n^k = 1, \\ e^{\alpha} u_n\left(\mathbf{P}_n^{k-1}\right), & \text{if } \lambda_n^k = 0, \end{cases} \tag{18}$$

where α is a positive parameter and $u_n(\mathbf{X}) \triangleq \mathbf{A}_n \mathbf{X} \mathbf{A}_n^T + \mathbf{Q}_n$ (see Algorithm 1).

To quantify the remote estimation quality of sensor n at time k, we define the trace of the expected RSE covariance for the sensor as follows:

$$\begin{aligned} J_n^k &= \mathrm{Tr}\left\{\mathbb{E}[\mathbf{P}_{n,r}^k]\right\} \\ &= \mathrm{Tr}\left\{\mathbb{P}\left[\gamma_n^k = 0\right] e^{\alpha} u_n\left(\mathbf{P}_n^{k-1}\right) + \mathbb{P}\left[\gamma_n^k = 1\right] e^{\alpha} \bar{\mathbf{P}}_n\right\} \end{aligned} \tag{19}$$

Based on the preceding system model and corresponding derivations, the optimization objective is to minimize the sum of the expected covariance of the RSE errors, which can be expressed as follows:

$$\text{Problem 1:} \min_{\theta^k, p_n^k, b_{n,g}^k} \frac{1}{K} \sum_{k=1}^{K} \sum_{n=1}^{N} J_n^k \tag{20}$$

$$\text{s.t.} \quad p_n^k \leq p_{\max}, \quad n \in \mathcal{N}, \tag{20a}$$

$$b_{n,g}^k \in \{0,1\}, \quad n \in \mathcal{N}, g \in \mathcal{G}, \tag{20b}$$

$$\sum_{g=1}^{G} b_{n,g}^k \leq I, \quad n \in \mathcal{N}, \tag{20c}$$

$$\sum_{n=1}^{N} p_n^k \|\mathbf{\Theta}_k \mathbf{h}_{n,k}^{sr}\|^2 + \|\mathbf{\Theta}_k\|^2 \sigma_{1,k}^2 \leq P_{\text{aRIS}}. \tag{20d}$$

where K is the maximum value of the time slot set. $P_{\text{aPIS}} = \xi(P_{\text{total}} - M(P_{\text{DC}} + P_c))$ represents the amplification power of the RIS [2]. Here, ξ is the amplifier efficiency, P_{total} is the total power budget, P_{DC} is the DC biasing power consumption, and P_c is the circuit power consumption of the active RIS. Constraint (20a) ensures that the transmission power of sensor n is within the maximum allowed limit. Constraints (20b) and (20c) enforce that each channel supports at most I sensors and that each sensor is assigned to only one channel. These two constraints together imply that the total number of sensors N must not exceed the overall channel capacity, i.e., $G \times I \geq N$, in order to ensure the feasibility of the sensor-to-channel assignment. For example, if $N = 15$ sensors are to be scheduled using $G = 3$ channels, and each channel can support up to $I = 5$ sensors, then the total capacity is $G \times I = 15$, and the assignment is feasible. However, if $N = 16$, at least one sensor cannot be accommodated, violating the constraints. Therefore, this relationship is essential to guarantee a valid channel allocation. Constraint (20d) limits the amplification power of the active RIS.

Algorithm 1 Iterative Algorithm Based on the KF Method.

Input: $\mathbf{A}_n, \mathbf{C}_n, \mathbf{Q}_n, \mathbf{R}_n, \mathbf{w}_n^k, \mathbf{v}_n^k, \epsilon, k_{\max}$.
Output: $\mathbf{P}_n^k$.
Initialization: $t = 1, \Delta = 1$.

while $\Delta > \epsilon$ **or** $t \leq t_{\max}$ **do**
  Calculate $P_{i,g}^t$ according to formulas (5)–(9)
  Update $\Delta = \mathbf{P}_n^{k|k} - \mathbf{P}_n^{k|k-1}$
  Update $k = k + 1$
end while
Obtain $\mathbf{P}_n^k = \bar{\mathbf{P}}_n^{k|k}$
for $k = 1$ to K **do**
  for $n = 1$ to N **do**
    for $g = 1$ to G **do**
      Set $b_{n,g}^k \in \{0,1\}$
      if $b_{n,g}^k = 1$ **then**
        $\mathbf{P}_n^k = \bar{\mathbf{P}}_n^{k|k}$
      else
        $\mathbf{P}_n^k = \mathbf{A}_n \left(\mathbf{P}_n^{k-1}\right) (\mathbf{A}_n)' + \mathbf{Q}_n$
      end if
    end for
  end for
end for

Based on the well-known properties of the standard KF (see Lemma 2.3 in [17]), it follows that $u_n\left(\mathbf{P}_n^{k-1}\right) \geq \bar{\mathbf{P}}_n$, where $\bar{\mathbf{P}}_n$ is a fixed constant. Consequently, by applying the natural logarithm function to (19), the expression can be expressed as

$$D_n^k = \ln(J_n^k) \approx \mathrm{Tr}\Big\{-\gamma_n^k + \alpha \ln\Big[u_n\Big(\mathbf{P}_n^{k-1}\Big) - \bar{\mathbf{P}}_n\Big]\Big\}. \tag{21}$$

Therefore, Problem 1 can be transformed into

$$\text{Problem 2: } \min_{\theta^k, p_n^k, b_{n,g}^k} \frac{1}{K}\sum_{k=1}^{K}\sum_{n=1}^{N}\Big(D_n^k\Big), \text{s.t. (20a), (20b), (20c), (20d).} \tag{22}$$

3. MADDPG-Based Resource Allocation

In this section, we first reformulate the optimization problem proposed in Problem 2 as a Multi-Agent Markov Decision Process (MDP) and then solve it using the MADDPG, Algorithm 2. As shown in Figure 2, the proposed system architecture follows a centralized training and decentralized execution (CTDE) framework, which is well suited to distributed wireless sensor networks. Specifically, the MADDPG algorithm is trained offline at a fusion center or an edge server, where both the actor and critic networks are optimized based on collected environmental data. After training, only the lightweight actor networks are deployed to individual sensor nodes, enabling each agent to independently make decisions based on its local observation without the need for online coordination. This ensures scalability and practicality in decentralized scenarios. Regarding the deployment of an active RIS, we assume it is installed at a fixed infrastructure location (e.g., a gateway node or controller hub), where an external or a battery-based power supply is available. Each active element is equipped with a low-power amplifier, and the number of elements is limited to ensure that the overall power consumption remains within practical bounds. The actions related to RIS beamforming are optimized jointly with the power control and sensor grouping through learned policies.

These implementation considerations ensure that the proposed active RIS-assisted MADDPG architecture is both theoretically grounded and practically feasible for deployment in low-power, distributed WSN environments.

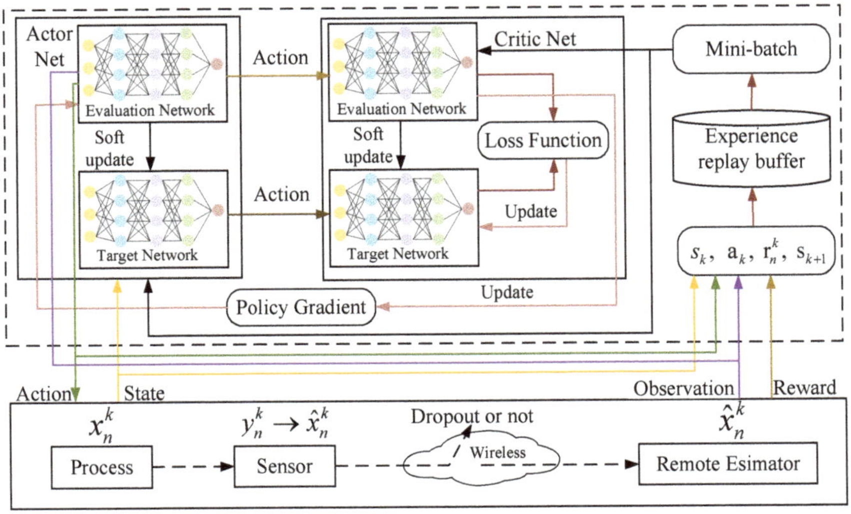

Figure 2. Schematic of MADDPG.

3.1. MDP Formulation

As depicted in Figure 2, each sensor functions as an individual agent. At time step k, agent n receives an observation $\mathbf{o_n^k}$ from the environment and takes an action $\mathbf{a_n^k}$. At time step k, given the set of states $\mathbf{s}^k = \{\mathbf{o_1^k}, \mathbf{o_2^k}, \ldots, \mathbf{o_N^k}\}$ and the set of joint actions $\mathbf{a}_k = \{\mathbf{a_1^k}, \mathbf{a_2^k}, \ldots, \mathbf{a_N^k}\}$, the agent receives an immediate reward r_n^k and the subsequent observation o_n^{k+1}. For sensor n, the observable information at time k comprises both the channel quality state based on its current location and the remote state estimation (RSE) error covariance. This can be represented as $\mathbf{o_n^k} = \{\mathbf{h}_n^k, \mathbf{P}_n^k\}$. The system state at time k is the collection of all sensor observations, expressed as $\mathbf{s_n^k} = \{\mathbf{o_n^k} \mid n \in \mathcal{N}\}$. At each time step, each sensor selects its action, which consists of the power allocation p_n^k, sensor grouping $b_{n,g}^k$, and beamforming selection for the RIS, denoted as $\mathbf{\Theta}^k$. Thus, the action $\mathbf{a_n^k}$ can be written as $\mathbf{a_n^k} = \{\mathbf{\Theta}^k, p_n^k, b_{n,g}^k\}$. The reward function assesses the agent's performance and guides its decision-making. The objective of this letter is to minimize the total RSE error of all sensors. MADDPG is typically used to solve maximization problems, where each agent aims to maximize its reward by selecting actions [18]. However, Problem 2 is a minimization problem, so we need to invert the objective value by defining the reward as the negative of the cost, i.e., $r_n^k = -D_n^k = \gamma_n^k - \alpha \ln\left[u_n\left(\mathbf{P}_n^{k-1}\right) - \bar{\mathbf{P}}_n\right]$.

3.2. The MADDPG Algorithm

In this section, we propose the RIS-MADDPG algorithm to address the optimization problem outlined above. The MADDPG algorithm is a widely used multi-agent reinforcement learning (MADRL) method for continuous control problems. For N agents, the MADDPG algorithm involves N actor networks and N critic networks, as shown in Figure 2. Each agent's actor network generates actions based on its policy, denoted as $\mu = \{\mu_1, \mu_2, \ldots, \mu_N\}$, with each policy implemented by a neural network $\phi^\mu = \{\phi_1^\mu, \phi_2^\mu, \ldots, \phi_N^\mu\}$. The critic networks, parameterized by $\phi^{\mathbf{Q}} = \{\phi_1^Q, \phi_2^Q, \ldots, \phi_N^Q\}$, evaluate the quality of the actions taken by the agents, considering the system state and the joint actions. Each agent n has an experience replay memory $\mathcal{B}_n$ that stores transition tuples $(\mathbf{s_k}, \mathbf{a}_k, r_n^k, \mathbf{s_{k+1}})$. At each time step, each agent samples a minibatch from its memory to update its policy. The gradient for agent n is computed as follows [18]:

$$\begin{aligned}
&\nabla_{\phi_n^\mu} \mathcal{J}(\mu_n) = \\
&\mathbb{E}_{\mathbf{a_n^k} \sim \mathcal{B}_n}\left[\nabla_{\phi_k^\mu} \mu_n\left(\mathbf{a_n^k} | \mathbf{o_n^k}\right) Q_n^\mu\left(\mathbf{s}^k, \mathbf{a}_k\right)|_{\mathbf{a_n^k} = \mu_n\left(\mathbf{o_n^k}\right)}\right].
\end{aligned} \tag{23}$$

To stabilize the learning process, MADDPG employs the target network technique. The target networks for the action and evaluation functions, μ_n' and Q_n', respectively, are updated using backpropagation. The loss function for updating the evaluation network is

$$\mathcal{L}\left(\phi_n^Q\right) = \mathbb{E}_{\mathbf{s}^k, \mathbf{a}_k, r_n^k, \mathbf{s}_{k+1}'}\left[\left(Q_n^\mu\left(\mathbf{s}^k, \mathbf{a}_k\right) - y_k\right)^2\right], \tag{24}$$

where $y_k = r_n^k + \psi Q_n'\left(\left(\mathbf{s}^k\right)', \mathbf{a}_k'\right)|_{\left(\mathbf{a_n^k}\right)' = \mu_n'\left(\mathbf{o_n^k}\right)}$.

After updating ϕ_n^μ and ϕ_n^Q, the parameters of the target networks, $\phi_n^{\mu'}$ and $\phi_n^{Q'}$, are softly updated as

$$\phi_n^{\mu'} = \tau_a \phi_n^\mu + (1 - \tau_a)\phi_n^{\mu'}, \tag{25}$$

$$\phi_n^{Q'} = \tau_c \phi_n^Q + (1 - \tau_c)\phi_n^{Q'}, \tag{26}$$

where τ_a and τ_c are the update parameters for the target network, and $\mathcal{C}$ denotes the update interval that determines how frequently the actor and target networks are updated during training. The detailed steps of the RIS-MADDPG algorithm are provided in Algorithm 2.

Algorithm 2 RIS-MADDPG Algorithm.

Initialize the actor network $\mu_n(\mathbf{o_n^k};\theta_n^{\mu})$ and the critic network $Q_n(\mathbf{s_n^k};\theta_n^{Q})$ for each agent n at time step k;
Initialize the target network $\mu_n'(\mathbf{o_n^k};\theta_n^{\mu'})$ and the critic nework $Q_n'(\mathbf{s_n^k};\theta_n^{Q'})$ for each agent n at time step k;
Initialize the experience replay buffer $\mathcal{B}_n$ for each agent n;
for episode = 1 to E **do**
  Initialize state $\mathbf{s}^k = \{\mathbf{o_1^k}, \mathbf{o_2^k}, \ldots, \mathbf{o_N^k}\}$ and $k = 1$
  while step $k < K$ **do**
    for each agent n **do**
      Get o_n^k, select action $\mathbf{a_n^k} = \mu_n(\mathbf{o_n^k})$
      Execute actions $\mathbf{a}_k$ and obtain the reward r_n^k and the next state $\mathbf{o_n^{k+1}}$
      Store $(\mathbf{s_k}, \mathbf{a}_k, r_n^k, \mathbf{s_{k+1}})$ in $\mathcal{B}_n$
    end for
  end while
  for each agent n **do**
    Sample a random mini-batch of $\mathcal{S}$ samples from the experience replay buffer $\mathcal{B}_n$
    Update the critic network by minimizing the loss function $\mathcal{L}(\phi_n^Q)$ in (24)
    if $k \mod \mathcal{C} = 0$ **then**
      Update the parameters ϕ_n^Q using the deterministic policy gradient in (23)
      Update two target networks: (25) and (26)
    end if
  end for
end for

To ensure the reproducibility of our results, all of the hyperparameters, training configurations, and simulation environment details have been clearly specified in Table 1, and the algorithmic process has been fully described in Algorithm 2.

Table 1. Simulation parameters.

Description	Parameter and Value
Transmission model	N = 9, G = I = 3, M = 16, $p_{\max}$ = 20 dBm $\delta_R = \delta_R = 4$ P_{aRIS} = 32 dBm P_{tot} = 30 dBm, P_{DC} = −5 dBm, P_c = −10 dBm $\alpha_R = 2.7$, $\alpha_B = 1.5$, $\alpha_S = 1.6$, $\alpha_m \in [1.1, 2]$ $\xi = 0.8$, $\alpha = 10^{17}$, σ_1^2 = −70 dBm, σ_2^2 = −80 dBm
KF model	$\mathbf{A}_n, \mathbf{C}_n, \mathbf{Q}_n, \mathbf{R}_n \in [0.5, 1.5]$
MADDPG	E = 10, K = 4000, $\mathcal{B}_n$ = 128, $\mathcal{C}$ = 2000

3.3. Computational Complexity Analysis

The computational complexity of Algorithm 1 mainly arises from two iterative phases: the convergence phase for updating the estimation covariance $\mathbf{P}_n^{k|k}$ and the scheduling phase involving nested loops over time slots (K), agents (N), and groups (G). Specifically, the convergence phase's complexity is $\mathcal{O}(k_{\max} \times N \times G)$, and the scheduling phase's complexity is $\mathcal{O}(K \times N \times G)$. Thus, the total complexity of Algorithm 1 is $\mathcal{O}(N \times G(k_{\max} + K))$, reflecting linear scalability in terms of the agents and groups.

For the MADDPG algorithm used to solve Problem 2, the complexity consists of two parts: centralized critic network training and decentralized actor network execution.

The centralized training's complexity is $\mathcal{O}(E^{\max} \times N)$, where $E^{\max}$ is the maximum number of training episodes. The decentralized execution complexity per agent is constant per episode, resulting in an overall execution complexity of $\mathcal{O}(E^{\max})$. Hence, the total complexity is $\mathcal{O}(E^{\max} \times N)$. Compared to centralized approaches like DDPG (with complexity $\mathcal{O}(E^{\max} \times N^2)$), MADDPG significantly reduces the computational overhead, thus enhancing the scalability and efficiency for multi-agent scenarios.

4. Simulation Results

Based on the above discussion, we divide the computational complexity of applying the MADDPG algorithm to solving Problem 2 into three components:

- The critic network (centralized training): During the centralized training phase, the critic network evaluates the global state–action pairs involving all N agents. Suppose that the maximum number of episodes is denoted as $E^{\max}$, and each agent requires a complexity of $\mathcal{O}(1)$ for a single update: the overall complexity becomes $\mathcal{O}(E^{\max}N)$.
- The actor network (decentralized execution): In the decentralized execution phase, each agent independently selects actions using its actor network. Since each agent makes decisions independently with the complexity $\mathcal{O}(1)$ per episode, the total complexity for all episodes is $\mathcal{O}(E^{\max})$.

Therefore, the total computational complexity of the MADDPG algorithm is represented as

$$\mathcal{O}(E^{\max}N + E^{\max}) = \mathcal{O}(E^{\max}N) \tag{27}$$

Moreover, we critically analyze the overhead costs compared to those of the baseline models. Specifically, compared to the centralized DDPG method—which incurs a complexity of approximately $\mathcal{O}(E^{\max}N^2)$ due to the joint optimization of the actions over all agents—MADDPG significantly reduces the complexity by decentralizing the actor network computations, compared to that of the standard reinforcement learning methods.

Figure 3 compares our proposed MADDPG algorithm with active RIS assistance against three baseline approaches: MADDPG with a passive RIS, DDPG with an active RIS, and MADDPG without RIS assistance. The results demonstrate substantial reductions in the performance errors provided by the active RIS. Specifically, the MADDPG algorithm with active RIS assistance achieves an 88% reduction in error compared to that with a passive RIS and a 200% reduction compared to that with no RIS assistance. Additionally, the active RIS within the MADDPG framework surpasses active-RIS-assisted DDPG by approximately 87% in its error reduction. These results clearly highlight the superior performance of the MADDPG algorithm with the active RIS, underscoring the potential of the active RIS to optimize WSNs. To enhance the clarity further, we have specified in the figure captions that the horizontal axis denotes the iteration steps and the vertical axis corresponds to the normalized estimation errors, both of which are dimensionless quantities.

Figure 4 illustrates the impact of the number of reflection elements on the sensor's RSE error. It can be observed that increasing the number of elements to 128 results in substantial error reductions of 32%, 78%, and 120% when compared to configurations with 32, 16, and 8 elements, respectively. This performance gain is attributed to the improved signal reflection and enhanced spatial diversity enabled by more reflection elements. These results demonstrate that a higher number of elements significantly enhances the sensing accuracy, thereby playing a crucial role in reducing the estimation error.

Figure 5 presents the convergence behavior of the RSE error under different maximum power constraints (10 dBm, 15 dBm, and 20 dBm). While the general convergence trends are visually similar, higher power constraints clearly lead to faster convergence and a lower steady-state RSE error. This is because larger transmission power budgets improve the signal-to-noise ratio (SNR) at the receiver, enabling more precise state estimations. These findings indicate that the power allocation not only influences the final estimation accuracy but also affects the convergence speed, highlighting its critical importance in dynamic WSN environments. In summary, the results in Figures 4 and 5 jointly confirm that both the number of reflection elements and the power constraints have a pronounced impact on the sensing performance. Optimizing these parameters leads to more accurate and responsive sensing, which is essential for reliable state estimation in resource-constrained wireless sensor networks.

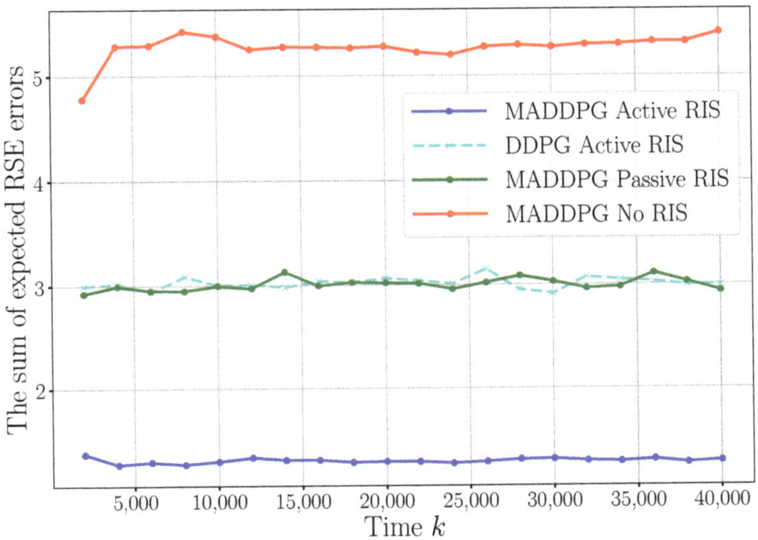

Figure 3. The sum of expected RSE errors under different algorithms.

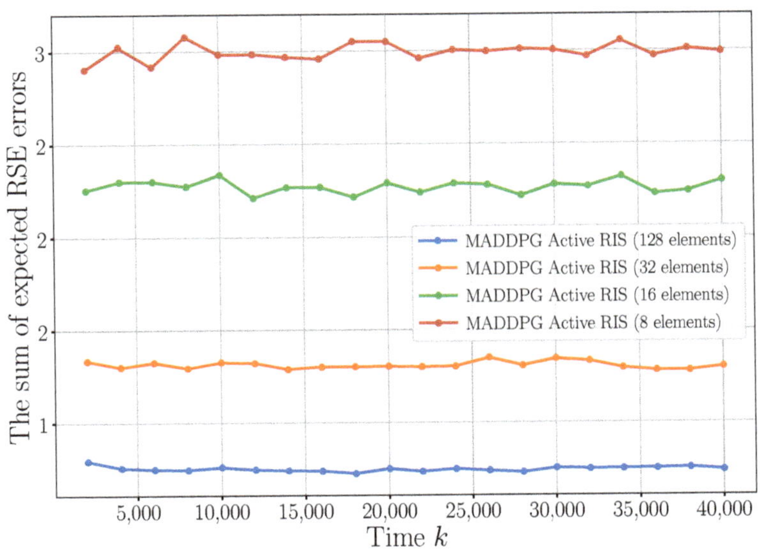

Figure 4. The convergence for different numbers of RIS reflecting elements.

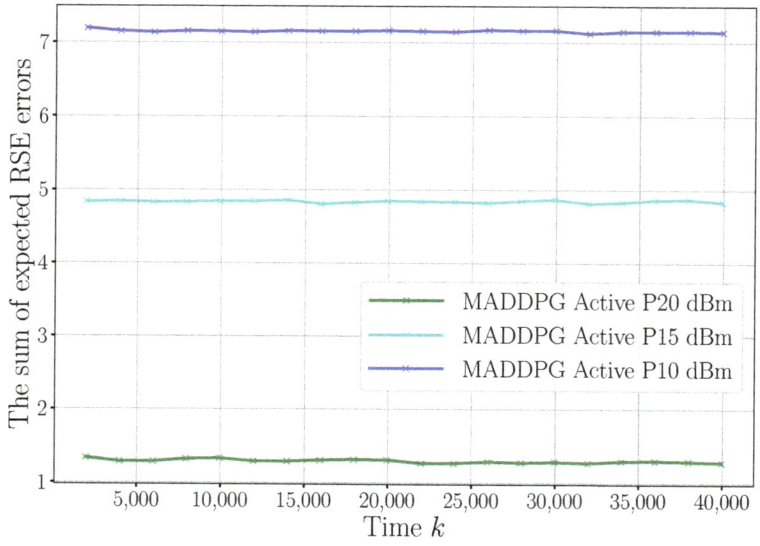

Figure 5. Convergence for different maximum power constraints.

To investigate the system-level trade-offs further, Figure 6 presents the convergence behavior of the proposed algorithm under different sensor-to-channel configurations (e.g., two sensors with six channels, three sensors with four channels, and four sensors with three channels). It can be observed that as the number of sensors increases under limited channel resources, the RSE error grows due to heightened interference and limited degrees of freedom for resource allocation. This validates the system's sensitivity to the sensor density and available spectrum.

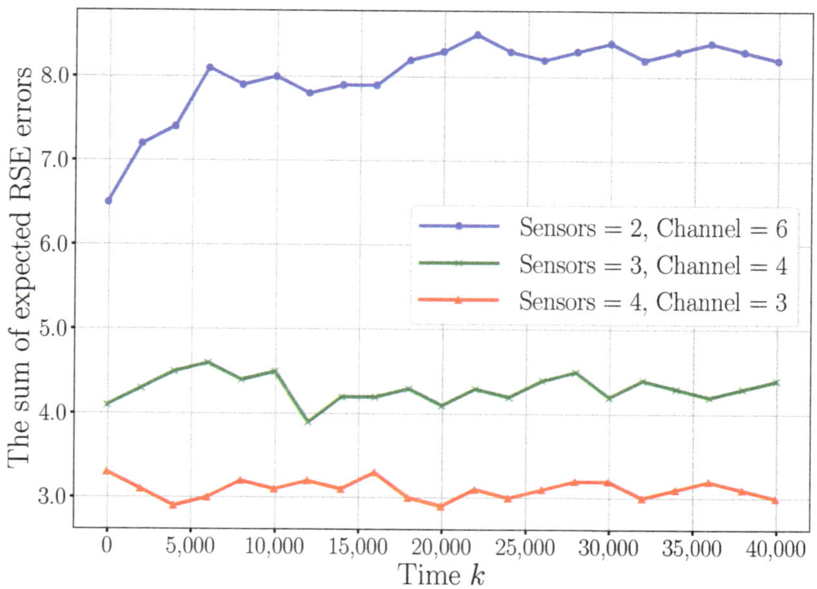

Figure 6. Convergence comparison under different sensor-to-channel configurations.

Figure 7 evaluates the impact of different amplifier efficiency coefficients ξ on the convergence behavior of the proposed algorithm. The parameter ξ reflects the effective amplification capability of the active RIS, as constrained by the total power budget P_{total} and the internal power consumption $M(P_{\text{DC}} + P_c)$, as defined in constraint (20d). A larger value for ξ enables the active RIS to contribute more effectively to signal enhancement via stronger amplification, thereby improving the signal-to-noise ratio (SNR) at the receiver and reducing the RSE error. However, increasing ξ also implies that more power is allocated to RIS amplification, potentially limiting the power available for sensors. As shown in the figure, a moderate value for ξ achieves a good trade-off between RIS enhancement and the overall system power balance, whereas excessive amplification may lead to diminishing returns. This result highlights the importance of careful tuning of the RIS-related parameters to optimize the joint communication–sensing performance in energy-constrained WSN environments.

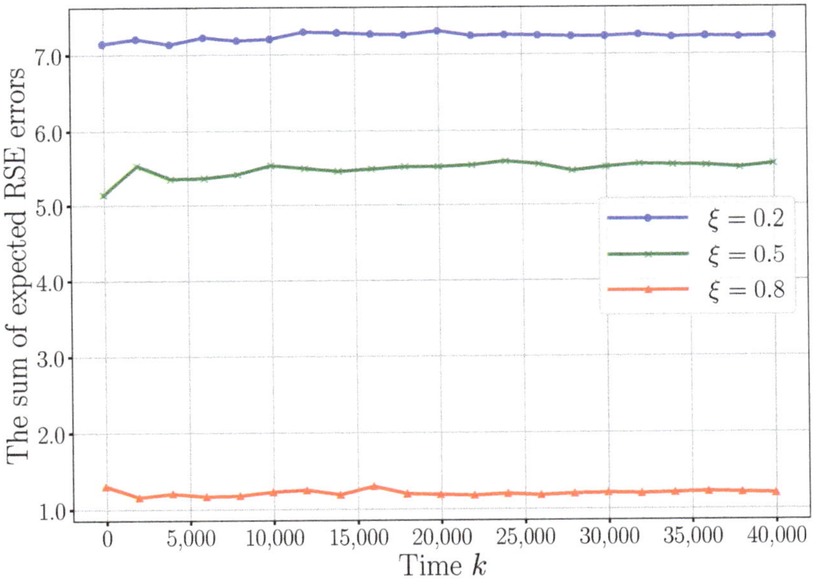

Figure 7. Convergence performance under different RIS amplifier efficiency coefficients ξ.

To facilitate a comprehensive and intuitive comparison across different experimental settings, Table 2 presents a summary of the convergence time and the steady-state RSE error under various configurations. In this table, "Conv. Time" denotes the number of iterations (abbreviated as "Iter.") required for the algorithm to reach convergence, while "Conv. Value" represents the final RSE error after stabilization. The data in the table are systematically derived from key experimental results, primarily based on the quantitative trends observed in Figures 3–5. As shown in Table 2, the proposed MADDPG algorithm assisted by the active RIS consistently achieves a superior performance in terms of both the convergence speed and estimation accuracy, particularly when equipped with a larger number of reflecting elements. In contrast, reducing the number of RIS elements or the transmission power constraint leads to increased estimation errors, highlighting the importance of spatial diversity and power resources for accurate sensing. This table provides a clear and concise overview of the system behavior across algorithms and configurations, offering valuable insights for future research and parameter optimization in wireless sensor networks.

Table 2. A summary of the convergence times and convergence values under different configurations.

Scenario	Conv. Time (Iter.)	Conv. Value
MADDPG + Active RIS (128 elements, 20 dBm)	3500	0.75
MADDPG + Active RIS (32 elements, 20 dBm)	3300	1.25–1.35
MADDPG + Active RIS (16 elements, 20 dBm)	3200	2.20–2.35
MADDPG + Active RIS (8 elements, 20 dBm)	3000	3.00
MADDPG + Active RIS (16 elements, 10 dBm)	3500	7.10–7.25
MADDPG + Active RIS (16 elements, 15 dBm)	3500	4.85–5.00
DDPG + Active RIS (16 elements, 20 dBm)	4500	3.10–3.20
MADDPG + Passive RIS (16 elements, 20 dBm)	5000	3.00
MADDPG without RIS (16 elements, 20 dBm)	10,000	5.15–5.25

5. Conclusions

This letter proposes an active RIS-assisted uplink NOMA technique using the MADDPG framework, aiming to minimize the sum of the expected RSE errors during wireless transmission. By leveraging the capabilities of an active RIS and the multi-agent learning capability of MADDPG, the proposed method dynamically optimizes the resource allocation to enhance the sensing and transmission performance in wireless networks. The simulation results validate that the proposed algorithm achieves notable improvements in minimizing the sum of the RSE errors, outperforming the conventional algorithms by approximately 87% to 200%. These results demonstrate the effectiveness of combining intelligent reflecting surfaces with reinforcement learning for joint communication and sensing design. Despite these promising results, the proposed approach also has several limitations. The offline training phase of MADDPG can be computationally intensive, especially with many agents, and the convergence behavior may be sensitive to hyperparameter tuning. In future work, we plan to explore more lightweight and scalable learning strategies while also evaluating additional performance metrics such as the convergence speed, latency, and energy efficiency. Moreover, extending the framework to support mobile agents and distributed multi-cell deployments will enhance its applicability further in real-world WSN scenarios.

Author Contributions: Conceptualization, R.L. and L.X.; methodology, R.L. and L.X.; software, R.L.; validation, L.X.; formal analysis, R.L. and L.X.; investigation, R.L.; resources, R.L.; data curation, R.L.; writing—original draft preparation, R.L.; writing—review and editing, R.L. and L.X.; visualization, R.L. and L.X.; supervision, R.L.; project administration, L.X. All authors have read and agreed to the published version of the manuscript.

Funding: This research received no external funding.

Institutional Review Board Statement: Not applicable.

Informed Consent Statement: Not applicable.

Data Availability Statement: The data that support the findings of this study are available from the corresponding author upon reasonable request.

Conflicts of Interest: The authors declare no conflict of interest.

References

1. Sharma, N.; Gautam, S.; Chatzinotas, S. Fractional Programming Based Optimization Techniques for RIS-Assisted SWIPT-IoT System. *IEEE Commun. Lett.* **2024**, *28*, 2819–2823. [CrossRef]
2. Peng, Z.; Weng, R.; Zhang, Z.; Pan, C.; Wang, J. Active Reconfigurable Intelligent Surface for Mobile Edge Computing. *IEEE Wirel. Commun. Lett.* **2022**, *11*, 2482–2486. [CrossRef]
3. Li, Y.; Mehr, A.S.; Chen, T. Multi-sensor transmission power control for remote estimation through a SINR-based communication channel. *Automatica* **2019**, *101*, 78–86. [CrossRef]

4. Rito, B.Y.D.; Li, K.H. RIS-Assisted NOMA With Interference-Cancellation and Interference-Alignment Scheme. *IEEE Commun. Lett.* **2023**, *27*, 1011–1015. [CrossRef]
5. Chai, F.; Zhang, Q.; Yao, H.; Xin, X.; Wang, F.; Xu, M.; Xiong, Z.; Niyato, D. Multi-Agent DDPG Based Resource Allocation in NOMA-Enabled Satellite IoT. *IEEE Trans. Commun.* **2024**, *72*, 6287–6300. [CrossRef]
6. Xu, X.; Wu, W.; Song, R. Transmit Power Minimization for RIS-Assisted MIMO-NOMA System. *IEEE Commun. Lett.* **2024**, *28*, 1367–1371. [CrossRef]
7. Li, J.; Yang, L.; Wu, Q.; Lei, X.; Zhou, F.; Shu, F.; Mu, X.; Liu, Y.; Fan, P. Active RIS-Aided NOMA-Enabled Space- Air-Ground Integrated Networks With Cognitive Radio. *IEEE J. Sel. Areas Commun.* **2025**, *43*, 314–333. [CrossRef]
8. Kumar, S.; Kumbhani, B.; Darshi, S. Performance Analysis of STAR-RIS Aided Short-Packet NOMA Network Under Imperfect SIC and CSI. *IEEE Syst. J.* **2025**, *19*, 600–611. [CrossRef]
9. Li, X.; Pei, Y.; Yue, X.; Liu, Y.; Ding, Z. Secure Communication of Active RIS Assisted NOMA Networks. *IEEE Trans. Wirel. Commun.* **2024**, *23*, 4489–4503. [CrossRef]
10. Xu, S.; Wu, S.; Zheng, Y. Joint dynamic user pairing and resource allocation for RIS-aided multi-cluster NOMA networks. *Phys. Commun.* **2025**, *68*, 102595. [CrossRef]
11. Wang, H.; Xu, L.; Yan, Z.; Gulliver, T.A. Low-Complexity MIMO-FBMC Sparse Channel Parameter Estimation for Industrial Big Data Communications. *IEEE Trans. Ind. Inform.* **2021**, *17*, 3422–3430. [CrossRef]
12. Wu, L.; Qu, J.; Li, S.; Zhang, C.; Du, J.; Sun, X.; Zhou, J. Attention-Augmented MADDPG in NOMA-Based Vehicular Mobile Edge Computational Offloading. *IEEE Internet Things J.* **2024**, *11*, 15. [CrossRef]
13. Khalid, W. DRL-Based Approach for RIS-Aided NOMA System in Short Packet Communications. In Proceedings of the 2025 8th International Conference on Circuits, Systems and Simulation (ICCSS), Ho Chi Minh City, Vietnam, 16–18 May 2025; pp. 90–93. [CrossRef]
14. Pang, G.; Liu, W.; Li, Y.; Vucetic, B. DRL-Based Resource Allocation in Remote State Estimation. *IEEE Trans. Wirel. Commun.* **2023**, *22*, 4434–4448. [CrossRef]
15. Li, Y.; Quevedo, D.E.; Lau, V.; Shi, L. Online sensor transmission power schedule for remote state estimation. In Proceedings of the 52nd IEEE Conference on Decision and Control, Firenze, Italy, 10–13 December 2013; pp. 4000–4005. [CrossRef]
16. Yu, J.; Li, Y.; Liu, X.; Sun, B.; Wu, Y.; Hin-Kwok Tsang, D. IRS Assisted NOMA Aided Mobile Edge Computing With Queue Stability: Heterogeneous Multi-Agent Reinforcement Learning. *IEEE Trans. Wirel. Commun.* **2023**, *22*, 4296–4312. [CrossRef]
17. Shi, L.; Johansson, K.H.; Qiu, L. Time and Event-based Sensor Scheduling for Networks with Limited Communication Resources. *IFAC Proc. Vol.* **2011**, *44*, 13263–13268. [CrossRef]
18. Wang, S.; Song, X.; Song, T. Joint Computation Offloading with Phase-Shift Design for RIS-Assisted Multi-UAV MEC Network. In Proceedings of the 2024 IEEE 99th Vehicular Technology Conference (VTC2024-Spring), Singapore, 24–27 June 2024; pp. 1–7. [CrossRef]

sensors

MDPI

Article

Joint Optimization of Radio and Computational Resource Allocation in Uplink NOMA-Based Remote State Estimation

Rongzhen Li and Lei Xu *

School of Computer Science and Engineering, Nanjing University of Science and Technology, Nanjing 210094, China; lirongzhen0213@163.com
* Correspondence: xulei_marcus@126.com

Abstract: In industrial wireless networks beyond 5G and toward 6G, combining uplink non-orthogonal multiple access (NOMA) with the Kalman filter (KF) effectively reduces interruption risks and transmission delays in remote state estimation. However, the complexity of wireless environments and concurrent multi-sensor transmissions introduce significant interference and latency, impairing the KF's ability to continuously obtain reliable observations. Meanwhile, existing remote state estimation systems typically rely on oversimplified wireless communication models, unable to adequately handle the dynamics and interference in realistic network scenarios. To address these limitations, this paper formulates a novel dynamic wireless resource allocation problem as a mixed-integer nonlinear programming (MINLP) model. By jointly optimizing sensor grouping and power allocation—considering sensor available power and outage probability constraints—the proposed scheme minimizes both estimation outage and transmission delay. Simulation results demonstrate that, compared to conventional approaches, our method significantly improves transmission reliability and KF estimation performance, thus providing robust technical support for remote state estimation in next-generation industrial wireless networks.

Keywords: uplink NOMA; outage risk; remote state estimation; coalitioin game; dinkelbach method; successive convex approximation method; dual decomposition method

1. Introduction

The rapid advancements in beyond 5G and 6G technologies have dramatically improved key aspects of wireless communication, such as bandwidth, latency, and connection density, thereby creating new opportunities for the development of wireless sensor networks (WSNs) [1–3]. However, these advancements also introduce new challenges, particularly in the efficient management of resources due to limited power, bandwidth constraints, and the complex interactions between critical system components, such as sensor grouping, power allocation, and state estimation. non-orthogonal multiple access (NOMA), a promising approach for resource sharing, leverages power-domain differentiation to improve system performance, particularly in the uplink. At the same time, remote state estimation, often optimized using the Kalman filter (KF), is crucial for accurately estimating a system's state, especially under noisy conditions in wireless networks. Despite the potential of both technologies, integrating NOMA with KF-based remote state estimation presents unique challenges, including interference management, power control, transmission delay, and outage risks. This paper proposes a joint optimization framework that combines NOMA's resource allocation capabilities with KF for remote state estimation, addressing these challenges simultaneously. By optimizing sensor grouping, power allocation, and transmission strategies, this approach minimizes outage risks while enhancing both the accuracy of remote state estimation and transmission efficiency, ensuring better system performance and sustainability in beyond 5G and 6G technologies.

During the transmission of sensor local states to the base station, limitations in bandwidth, power, and other resources, as well as factors such as delay, packet loss, and channel interference, make it a critical issue to consider how to assign sensors to channels and set appropriate power levels for each sensor to minimize the total outage risk across all sensors. To address this, we propose combining game theory with sensor grouping and power allocation strategies in NOMA as an effective approach [4–6]. For instance, Liu et al. propose a greedy subchannel matching algorithm based on hypergame theory for sensor and subchannel pairing in downlink NOMA networks, highlighting the importance of sensor grouping in WSN [7]. Additionally, a game theory-driven data security transmission method is proposed to mitigate potential interference in drone-assisted vehicular networks, with the significant impact of mutual interference on communication quality being demonstrated [8]. Wang et al. explore the application of game theory in cooperative communication scenarios, proposing a two-stage game model based on power and spectrum trading, optimizing transmission rates by solving the game equilibrium, and emphasizing the key role of power in communication processes [9,10].

While some models in sensor wireless transmission are convex functions, many others are non-convex, which makes traditional game theory methods ineffective in solving such problems. As a result, researchers have proposed solutions to convert non-convex problems into convex optimization problems [11–13]. For example, in reference [14], a joint optimization algorithm that combines the Dinkelbach algorithm with a successive convex approximation (SCA) method successfully converts fractional optimization problems into equivalent linear problems, thus converting non-convex problems into convex ones. This method proves the existence of an optimal solution for the joint optimization of power and subcarrier allocation. The complex constrained optimization problems are simplified using Lagrangian dual decomposition in [15,16], where power allocation in uplink NOMA systems is optimized, leading to a significant reduction in computational complexity. Transmission delay is another critical factor. Transmission delay, energy consumption, and sensor grouping are jointly optimized in [17,18], where their impacts on data transmission are studied, emphasizing the indispensable role of transmission delay in the optimization process.

Although existing research provides valuable insights into addressing some problems in sensor wireless transmission, several challenges remain unsolved. First, effectively integrating transmission delay, energy consumption, channel interference, and outage risk due to remote state estimation in beyond 5G and 6G environments remains a complex problem. Second, while some methods have addressed parts of non-convex optimization problems, further research is needed to reduce system complexity while maintaining computational efficiency. Additionally, current game theory models primarily focus on single-objective optimization; thus, balancing multiple objectives in a multi-objective optimization framework requires further exploration.

Minimizing outage risks for sensors during transmission is critical for ensuring the stability and reliability of uplink NOMA systems [19–22]. Yadav et al. introduce the optimization of intelligent reflecting surface reflection coefficients, which effectively reduces the impact of channel fading and lowers outage probability [23]. However, it does not fully address the effects of power allocation and non-ideal intelligent reflecting surfaces on the system. Power allocation and sensor grouping are optimized to reduce outage risks and increase system capacity, though the study lacks a detailed discussion of interference management and dynamic resource scheduling strategies. In [24], outage probability in downlink NOMA systems is minimized through energy harvesting and cooperative communication optimization; however, the paper does not explore how to further improve performance via enhanced CSI estimation and optimized power allocation. The optimization of power allocation is used to enhance system rate and reduce outage probability, though the specifics of sensor grouping and interference management are not delved into in the study [25]. Overall, while these methods effectively reduce outage probability, they do not fully explore refined resource scheduling strategies. Future research could investigate the impact of these factors on system performance.

When addressing the minimization of outage risk during the remote state estimation process for sensors, we must not only consider sensor state estimation issues (e.g., the state estimation error covariance) but also focus on the communication challenges in beyond 5G and 6G wireless transmission. Unlike conventional studies that rely purely on quality-of-service (QoS) metrics or signal-to-interference-plus-noise ratio (SINR) for sensor grouping and power allocation, this work focuses explicitly on integrating remote state estimation, network communication processes, and outage probabilities during transmission. Specifically, the complexity arises from the necessity to balance multiple performance criteria—remote state estimation accuracy, transmission latency, and outage probability—in a unified optimization framework, which underscores the importance of jointly utilizing uplink NOMA and KF techniques. The main contributions of this paper are summarized as follows

1. Problem Modeling: We propose a novel unified optimization framework, formulating the joint minimization of outage risk in remote state estimation and transmission delay as a mixed-integer nonlinear programming (MINLP) problem. This formulation explicitly incorporates both communication and estimation constraints, providing a theoretical foundation for subsequent algorithm designs.
2. Coalitional Game-Based Sensor Grouping: Motivated by the limitations of traditional heuristic approaches, we propose a coalitional game-based sensor grouping algorithm. In this framework, each sensor acts as a rational player seeking to maximize its individual utility, and sensors iteratively form cooperative groups based on local improvement rules. The game converges to a Nash-stable coalition structure, which ensures a locally optimal grouping strategy. This approach effectively mitigates inter-group interference in multi-sensor uplink scenarios while significantly enhancing grouping efficiency and resource utilization.
3. Power Allocation Optimization: To address the nonlinearity and fractional nature of the power allocation subproblem, we employ the Dinkelbach algorithm to transform the original fractional objective into a parameterized form. Then, we use successive convex approximation (SCA) to handle non-convex constraints and apply dual decomposition to solve the resulting convex subproblems. This layered solution framework guarantees convergence to a stationary point and achieves a suboptimal solution with provable performance. Simulation results confirm that this approach provides near-optimal performance with significantly reduced computational complexity.

By combining these strategies, our framework effectively mitigates outage risks during sensor transmissions and considerably enhances overall system reliability and estimation performance. Additionally, by transforming the original non-convex optimization into convex subproblems, the proposed solution methodology substantially reduces computational complexity. These contributions provide theoretical insights and practical guidelines for optimizing wireless transmission in uplink NOMA systems within future industrial wireless networks beyond 5G and 6G.

To further highlight the novelty of our approach, we provide a comparison in Table 1 that contrasts the proposed framework with representative existing methods in terms of sensor grouping, power allocation, outage-awareness, and estimation integration.

The remainder of this paper is organized as follows: Section 2 presents the mathematical formulation of the problem, Section 3 investigates the coalition game-based sensor grouping problem and the joint optimization algorithm for solving the optimal power allocation, while numerical examples and conclusions are presented in Sections 4 and 5, respectively.

Table 1. Comparison of the proposed method with existing approaches.

Feature	In [26]	In [22]	Proposed Method
Sensor grouping	Not considered	Heuristic clustering	Game-theoretic
Power allocation	Fixed allocation	Cognitive + Fairness	Dinkelbach + SCA
Outage	Not addressed	Closed-form	Closed-form
Estimation	KF	Ignored	KF

Notations: $\mathbb{R}$ denotes the set of real numbers. $\mathbb{C}$ represents the set of complex numbers. The symbol $(\cdot)^T$ denotes the transpose operation. The superscript $(\cdot)^{-1}$ indicates the matrix inverse. The symbol $(\cdot)'$ represents the vector or matrix transpose operator. The notation $\mathbb{P}[\cdot]$ refers to probability. $\mathrm{Tr}(\cdot)$ represents the trace operator of a matrix.

2. System Model

This paper investigates an uplink NOMA transmission model in a wireless sensor network with a single base station. As illustrated in Figure 1, after performing local processing, N sensors transmit their data to the base station, which subsequently conducts remote state estimation. Before transmission, each sensor independently evaluates and selects one of the G channels for sensor grouping based on a comprehensive assessment of factors such as its distance to the base station, channel gain, maximum and minimum power constraints, and noise levels. Additionally, within each group, the decoding order and power allocation strategy are determined to optimize transmission performance and resource utilization efficiency. This process highlights the sensors' adaptive capabilities in channel selection, grouping, and resource scheduling, which are critical components of the system design. The set of cellular sensors is denoted by $\mathcal{N} = \{1, 2, \dots, N\}$, with each sensor equipped with a single antenna. In an uplink NOMA system for beyond 5G and 6G networks, the same channel can be assigned to multiple sensors. It is assumed that all sensors in the base station share $\mathcal{G}$ channels, resulting in $\mathcal{G} = \{1, 2, \dots, G\}$ NOMA clusters within the cell. The set of sensor indices in each cluster is defined as $\mathcal{I}_g = \{1, 2, \dots, I_g\}$, where I_g denotes the number of sensors assigned to the g-th NOMA cluster. It should be noted that the set $\mathcal{I}_g$ may vary across different clusters depending on the grouping results.

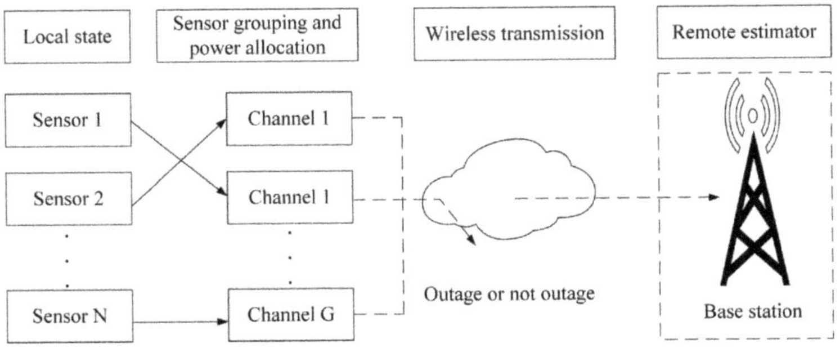

Figure 1. System model of remote state estimation outage risk.

2.1. Local State Estimate Model

To introduce the basic framework of the KF method, we describe how the measurement output of sensor i in channel g reflects the underlying process. The system's state at the next time step evolves from the previous state at time t according to the linear transition equation

$$x_{i,g}^{t+1} = A_i^g x_{i,g}^t + w_{i,g}^t, \tag{1}$$

where $x_{i,g}^t \in \mathbb{R}^{l_n}$, $w_{i,g}^t \in \mathbb{R}^{l_n}$, and $A_i^g \in \mathbb{R}^{l_n \times l_n}$ make up the state transition matrix. The state variable $x_{i,g}^{t+1}$ and the process noise $w_{i,g}^t$ are independent and follow complex Gaussian distributions. At the next time step, the observation is modeled by the linear measurement equation

$$y_{i,g}^{t+1} = C_i^g x_{i,g}^t + v_{i,g}^t, \tag{2}$$

where $y_{i,g}^{t+1} \in \mathbb{R}^{r_n}$ and $C_i^g \in \mathbb{R}^{r_n \times l_n}$ make up the measurement matrix and $v_{i,g}^t \in \mathbb{R}^{r_n}$ is the measurement noise, which also follows a complex Gaussian distribution [26,27]. The KF aims to use a sequence of observations to accurately estimate the state.

The prior and posterior state estimates, $\hat{x}_{i,g}^{t|t-1}$ and $\hat{x}_{i,g}^t$, respectively, are updated using the Kalman gain $K_{i,g}^t$ and the error covariances $P_{i,g}^{t|t-1}$ and $P_{i,g}^t$. The KF equations are as follows [26]:

$$\hat{x}_{i,g}^{t|t-1} = A_i^g \hat{x}_{i,g}^{t-1}, \tag{3}$$

$$P_{i,g}^{t|t-1} = A_i^g P_{i,g}^{t-1} (A_i^g)' + Q_i^g, \tag{4}$$

$$K_{i,g}^t = P_{i,g}^{t|t-1} (C_i^g)' \left[C_i^g P_{i,g}^{t|t-1} (C_i^g)' + R_i^g \right]^{-1}, \tag{5}$$

$$\hat{x}_{i,g}^t = \hat{x}_{i,g}^{t|t-1} + K_{i,g}^t \left(y_{i,g}^t - C_i^g \hat{x}_{i,g}^{t|t-1} \right), \tag{6}$$

$$P_{i,g}^t = \left(\nu - K_{i,g}^t C_i^g \right) P_{i,g}^{t|t-1}, \tag{7}$$

where ν is the identity matrix of size $l_n \times l_n$.

The estimation error covariance, denoted $P_{i,g}^t$, quantifies the uncertainty of the state estimate and is defined as

$$P_{i,g}^t \triangleq \mathbb{E}\left[\left(x_{i,g}^t - \hat{x}_{i,g}^t \right) \left(x_{i,g}^t - \hat{x}_{i,g}^t \right)' \right]. \tag{8}$$

As the local estimation error covariance $P_{i,g}^t$ converges exponentially to a steady-state value $\bar{P}_i^g$, the KF's accuracy improves over time.

2.2. Uplink NOMA Communication Model

The impulse response of the channel between sensor i and the base station is represented by $h_{i,g}^k = \frac{e_{i,g}^k}{\sqrt{1+\left(d_{i,g}^k\right)^\beta}}$, where $e_{i,g}^k$ denotes the impulse response of the Rayleigh fading channel, with β as the path loss exponent and $d_{i,g}^k$ as the distance between sensor i and the base station. This model captures both small-scale and large-scale effects, where the numerator models Rayleigh fading and the denominator models large-scale path loss. The square root form is used because path loss affects the signal power and, thus, the amplitude is scaled by the square root to ensure that the average received power decays as $1/(1+(d_{i,g}^k)^\beta)$. This ensures consistency with standard wireless propagation models. The probability density function of $|e_{i,g}^k|^2$ is given by $f_{|e_{i,g}^k|^2}(x) = \frac{1}{2\mu^2}e^{-\frac{x}{2\mu^2}}$, where μ is the variance of the normal distribution $N(0,\mu)$.

The received signal at the base station is given by

$$y_g^k = \sum_{i=1}^{I} h_{i,g}^k \sqrt{p_{i,g}^k} s_{i,g}^k + n_g, \tag{9}$$

where $p_{i,g}^k$ represents the transmit power allocated to sensor i in channel g and $s_{i,g}^k$ denotes the transmitted message from sensor i in channel g. Additionally, n_g represents the Gaussian white noise in the process of signal transmission from the base station to sensor i through channel g.

The probability density function (PDF) of $|e_i|^2$ [19] is given by

$$F_{\left|e_{i,g}^k\right|^2}(x) = \int_{-\infty}^{x} f_{\left|e_{i,g}^k\right|^2}(t)dt = 1 - e^{-\frac{x}{2\mu^2}}. \tag{10}$$

At the receiver in a NOMA system, SIC detection is applied to separate the multi-sensor superimposed signals. The optimal decoding order follows an ascending sequence based on sensors' channel gains, allowing for sequential detection and separation of each sensor's signal. Finally, the achievable transmission rate for sensor i in channel g can be expressed according to Shannon's theorem as follows: At the receiver in a NOMA system, SIC detection is applied to separate the multi-sensor superimposed signals. The optimal decoding order is determined by an ascending sequence of sensors' channel gains, thereby facilitating sequential detection and the separation of each sensor's signal. Finally, the achievable transmission rate for sensor i in channel g can be expressed according to Shannon's theorem as follows:

$$R_{i,g}^k = B\log_2(1+\mathrm{SINR}_{i,g}^k), \tag{11}$$

where B represents the bandwidth available to each sensor and $\mathrm{SINR}_{i,g}^k$ is defined as $\frac{p_{i,g}^k|h_{i,g}^k|^2}{\sum_{j=i+1}^{I} p_{j,g}^k|h_{j,g}^k|^2+\sigma^2}$.

The outage probability, as a critical metric for evaluating wireless system performance, represents the probability that the instantaneous link rate falls below the required sensor rate. We define $E_{i,g}^k$ as the event in which the base station can successfully decode the signals from sensors 1 to $i-1$ in channel g but fails to correctly decode the signal from sensor i in the same channel. When the sensor's rate requirement is $R_{i,g}^k$, the minimum transmission rate is $\hat{R}_{i,g}^k$ and $R_{i,g}^k < \hat{R}_{i,g}^k$; the outage probability of the event $E_{i,g}^k$ can be expressed as follows: [21]

$$\mathbb{P}(E_{i,g}^k) = \mathbb{P}\left\{R_{i,g}^k < \hat{R}_{i,g}^k\right\} = 1 - e^{-\frac{\phi_{i,g}^k}{2\mu^2}}, \tag{12}$$

where $\phi_{i,g}^k = \frac{(2^{\frac{\hat{R}_{i,g}^k}{B}}-1)(\sum_{j=i+1}^{I} p_{j,g}^k|h_{j,g}^k|^2+\sigma^2)\left(1+(d_{i,g}^k)^\beta\right)}{p_{i,g}^k}$. Correspondingly, this also represents its non-outage probability, which we define as $\mathbb{P}_c$. Based on the above derivation, the outage probability for sensor i in channel g is formulated as follows:

$$\mathbb{P}_{i,g}^{out,k} = 1 - \mathbb{P}_c(E_{1,g}^k \cap \cdots \cap E_{i,g}^k) = 1 - \prod_{z=1}^{i} e^{-\frac{\phi_{z,g}^k}{2\mu^2}}, \tag{13}$$

where

$$\phi_{z,g}^k = \begin{cases} \frac{(2^{\frac{R_{z,g}^k}{B}}-1)(\sum_{j=z+1}^{I} p_{j,g}^k |h_{j,g}^k|^2+\sigma^2)(1+(d_{z,g}^k)^\beta)}{p_{z,g}^k}, & z \neq I, \\ \frac{(2^{\frac{R_{z,g}^k}{B}}-1)(1+(d_{z,g}^k)^\beta)\sigma^2}{p_{z,g}^k}, & z = I. \end{cases}$$

2.3. *Remote State Estimation Model*

At each remote time k, all sensors must send their local estimates to the remote estimator over a shared wireless channel. We denote the transmission of $\hat{x}_{i,g}^k$ by a binary random process $\left\{\gamma_{i,g}^k\right\}$:

$$\gamma_{i,g}^k = \begin{cases} 1, & \text{if } \hat{x}_{i,g}^k \text{ arrives without errors at time } k, \\ 0, & \text{otherwise (regarded as dropout)}. \end{cases} \tag{14}$$

To explicitly reflect how packet dropouts influence the remote estimation process, we derive the expression of $P_{i,g}^{r,k}$ based on the binary transmission indicator $\gamma_{i,g}^k$ defined in (14). Specifically, when $\gamma_{i,g}^k = 1$, the remote estimator directly adopts the received local estimate and sets the estimation error covariance to the local value $\bar{P}_i^g$. Conversely, when $\gamma_{i,g}^k = 0$, the transmission fails, and the remote estimator performs a time prediction using the Kalman filter, resulting in the following recursive update:

$$P_{i,g}^{r,k} = \begin{cases} \bar{P}_i^g, & \text{if } \gamma_{i,g}^k = 1, \\ A_i^g\left(P_{i,g}^{r,k-1}\right)(A_i^g)' + Q_i^g, & \text{if } \gamma_{i,g}^k = 0. \end{cases} \tag{15}$$

Combining Equation (14) with Equation (13) from the previous section, we can draw the following conclusion:

$$\mathbb{P}\left[\gamma_{i,g}^k = 0\right] = 1 - \prod_{z=1}^{i} e^{-\frac{\phi_z^g}{2\mu^2}}. \tag{16}$$

2.4. *Transmission Delay Model*

2.4.1. *Local Transmission Time*

When sensor i in channel g performs its task locally, the task execution delay can be described by $T_{i,g}^{local} = \frac{I_i^g}{F^s}$, where F^s denotes the computational capability of the sensor.

2.4.2. *Base Station Offload and Processing Time*

If the sensor offloads its task to the base station, the total delay consists of the transmission delay over the wireless link and the execution delay at the base station, expressed as $T_{i,g}^{off} = T_{i,g}^{tra} + T_{i,g}^{base}$, where $T_{i,g}^{tra}$ and $T_{i,g}^{base}$ denote the transmission delay and the execution delay, respectively. The transmission delay is formulated as $T_{i,g}^{tra} = \frac{D_i^g}{R_i^g}$, and the execution delay at the base station is $T_{i,g}^{base} = \frac{I_i^g}{F^b}$, where F_b denotes the computational capability of the base station. The total time expenditure for the sensor is $T_{i,g}^{total} = T_{i,g}^{local} + T_{i,g}^{off}$.

2.5. *Problem Formulation*

In this section, we formulate an optimization problem to minimize the remote outage risk, which incorporates the total time expenditure and the associated error covariance of remote state estimation in KF. The objective is to jointly optimize sensor grouping and power allocation for the sensors. This objective balances transmission delay and estimation accuracy. To capture this trade-off, we introduce a weighted cost that accounts for both communication efficiency and estimation quality. Specifically, we define the system's objective value using the following equation:

$$
\begin{aligned}
\mathbf{J}_{i,g}^{k} = \mathrm{Tr}\{&(1-\mathbb{P}_{i,g}^{out,k})[\xi T_{i,g}^{total,k} + (1-\xi)\psi \bar{P}_{i}^{g}] \\
&+\mathbb{P}_{i,g}^{out,k}[\xi T_{i,g}^{local,k} + (1-\xi)\psi H_{i}^{g}(P_{i,g}^{r,k-1})]\},
\end{aligned}
\tag{17}
$$

where $H_i^g(X) \triangleq A_i^g X(A_i^g)' + Q_i^g$ and ξ (with $0 < \xi < 1$) make up a weight factor used to combine the objectives into a single function and ψ is a scaling factor to align the objectives to the same magnitude. The description of Problem 1 is as follows:

$$
\text{Problem 1:} \min_{p_{i,g}^{k}} \frac{1}{K}\sum_{k=1}^{K}\sum_{i=1}^{I}\sum_{g=1}^{G}\mathbf{J}_{i,g}^{k} \tag{18a}
$$

$$
\text{s.t.} \quad p_{i,g}^{k} \leq p_{\max}, \quad i \in \mathcal{I}, g \in \mathcal{G}, \tag{18b}
$$

$$
T_{i,g}^{total,k} \leq T_{i,g}^{max}, \quad i \in \mathcal{I}, g \in \mathcal{G}, \tag{18c}
$$

$$
\bigcup_{i\in\mathcal{I}}\bigcup_{g\in\mathcal{G}}\mathcal{U}_{i}^{g} = \mathcal{N}, \quad i \in \mathcal{I}, g \in \mathcal{G}, \tag{18d}
$$

$$
\mathcal{U}^{g} \cap \mathcal{U}^{g'} = \varnothing, \quad g, g' \in \mathcal{G}, \tag{18e}
$$

Constraint (18b) specifies that the transmission power of sensor i in channel g must be less than the maximum transmission power limits. Constraint (18c) ensures that the overall transmission time experienced by sensor i over channel g does not exceed the prescribed maximum threshold. Constraints (18d) and (18e) indicate that each sensor can be assigned to only one channel.

3. Solution of the Optimization Probelm

In this chapter, we propose an optimization method based on algorithm decomposition for the MINLP problem in Problem 2. Due to the high computational complexity typically associated with solving MINLP problems, we decompose the original problem into two subproblems to reduce the computational burden during the solution process, aiming to improve both the solving efficiency and practical feasibility of the implementation.

3.1. Grouping Strategy Based on Coalition Game Theory

To obtain the initial grouping $\mathcal{U}_{ini}$, sensors are uniformly and randomly assigned to each channel, achieving an even distribution of sensors across channels (see Algorithm 1).

In the above system model, the sensors are divided into multiple channels, each forming a coalition of the same size. Within each coalition, sensors collaboratively offload their respective data blocks to the base station, while competition exists between different coalitions. The objective of the optimization problem is to maximize the overall system performance. We define the outage risk under the joint optimization of remote state estimation and transmission delay as the utility function v within the coalition, representing the total outage risks of all sensors within the coalition, as follows:

$$
v(\mathcal{G}) = (\mathbb{O}_1, \mathbb{O}_2, \cdots, \mathbb{O}_N). \tag{19}
$$

Before the matching process, the mutual preference lists are indispensable. Both sensors and base station need to build their preference lists based on their own utilities. We define Γ as the total outage risks of all sensors within a coalition. For any sensors $\mathcal{U}_i^g \in \mathcal{N}$,

$\mathcal{U}_j^{g'} \in \mathcal{N}$ with $\mathcal{U}_i^g \in G_g$ and $\mathcal{U}_j^{g'} \in G_{g'}$, $G_g \succ_{\mathcal{U}_i^g \mathcal{U}_j^{g'}} G_{g'}$ indicates that sensor $\mathcal{U}_i^g$ prefers joining coalition G_g over $G_{g'}$. In this case, the sensor will exchange positions with those in $G_{g'}$. The final result is that after exchanging positions, the total outage risks of sensors in different coalitions is higher than before the exchange. The new partition can be denoted as $\Gamma\{\mathcal{G}'\} = \Gamma(G_g \setminus \{\mathcal{U}_i^g\} \cup \{\mathcal{U}_j^{g'}\}) + \Gamma(G_{g'} \setminus \{\mathcal{U}_j^{g'}\} \cup \{\mathcal{U}_i^g\})$.

Algorithm 1 Intelligent Sensor Grouping Algorithm Based on Coalition Game Theory under Fixed Power Constraints

Input: $\mathcal{N}, \mathcal{G}^0, \mathcal{I}, h_i^g, p_i^g, n \in \mathcal{N}, i \in \mathcal{I}, g \in \mathcal{G}^0$.
Output: Optimal grouping $\mathcal{G}_{fin}$.
Initialization: $t = 1,\ p_i^g = p_{\max},\ \mathcal{U}_{ini} = \mathcal{U}_i^g,\ \mathcal{G}^0 = \mathcal{G}_{ini},\ \forall i \in \mathcal{I},\ g \in \mathcal{G}^0$.

```
while 𝒢^t does not converge to Nash-stable partition G_fin do
    Select U_i^g ∈ G_g and U_j^{g'} ∈ G_{g'}, where g ≠ g', i ∈ I, j ∈ I
    Define 𝒢' = (𝒢^t \ {G_{g'}, G_g}) ∪ {G_{g'} \ {U_i^g} ∪ {U_j^{g'}}, G_g \ {U_j^{g'}} ∪ {U_i^{g'}}}
    Calculate E(G_m), E(G_{m'}) by Problem 4
    if Γ{𝒢'} > Γ{𝒢^t} then
        Update 𝒢^{t+1} = 𝒢'
        Update t = t + 1
    end if
end while
```

3.2. Power Allocation Based on Dinkelbach, SCA, and Dual Decomposition Methods

This section addresses the power allocation problem for a fixed sensor grouping by applying the Dinkelbach algorithm, along with the SCA and dual decomposition methods discussed below. To address the issue of offloading time consumption in the expectation, we use the Dinkelbach algorithm to handle the fractional programming. This method is particularly suitable because our objective function involves a ratio structure, and Dinkelbach's algorithm efficiently handles such problems with convergence guarantees. The transformed problem is described as follows (see Algorithm 2):

$$\mathbf{J}_{i,g}^k = \mathrm{Tr}\left\{ Z_{i,g}^k - \prod_{z=1}^{i} e^{-\frac{\phi_{z,g}^k}{2\mu^2}} \left(\lambda_{i,g}^k \frac{R_{i,g}^k}{\xi} + L_{i,g}^k \right) \right\}. \tag{20}$$

where $Z_{i,g}^k$ and $L_{i,g}^k$ represent $\xi T_{i,g}^{local,k} + (1-\xi)\psi \bar{P}_i^g$ and $(1-\xi)\psi Z_{i,g}^k - D_i^g - \xi T_{i,g}^{base,k}$, respectively, and $\lambda_{i,g}^k$ is a non-negative parameter. We define $F(\lambda_{i,g}^k) = -\min_{p_{i,g}^k}\left\{ \lambda_{i,g}^k \frac{R_{i,g}^k}{\xi} - D_i^g \right\}$.

Algorithm 2 Iterative Algorithm Based on Dinkelbach

Input: $\xi, D_i^g, R_{i,g}^k, \text{iter\_max}, \epsilon, i \in \mathcal{I}, g \in \mathcal{G}$.
Initialization: $\lambda_{i,g}^{\text{iter},k} = 0, \text{iter} = 1$.

```
while F(λ_{i,g}^{iter,k}) > ε or iter ≤ iter_max do
    F(λ_{i,g}^{iter,k}) = D_i^g − λ_{i,g}^{iter,k} R_{i,g}^k / ξ
    iter = iter + 1
    λ_{i,g}^{iter,k} = ξ D_i^g / R_{i,g}^k
end while
```

Output: Obtain $\lambda_{i,g}^k = \lambda_{i,g}^{\text{iter},k}$.

To facilitate subsequent analysis, we can derive from Equation (20) that the minimization problem above can be transformed into a maximization problem. Additionally, since $Z_i^g(k)$ is a constant, it has no impact on sensor grouping or power allocation. We redefine an objective function $\mathbf{O}_{i,g}^k$, which is derived from the original objective function $\mathbf{J}_{i,g}^k$ by excluding the $Z_{i,g}^k$ component and can be considered an approximation of $\mathbf{J}_{i,g}^k$.

$$\mathbf{O}_{i,g}^k \approx -\mathbf{J}_{i,g}^k = \mathrm{Tr}\left\{ \prod_{z=1}^{i} e^{-\frac{\phi_{z,g}^k}{2\mu^2}} \left(\lambda_{i,g}^k \frac{R_{i,g}^k}{\xi} + L_{i,g}^k \right) \right\}. \tag{21}$$

Therefore, Problem 1 can be reformulated as

$$\begin{aligned} &\text{Problem 2:} \max_{p_{i,g}^k} \frac{1}{K} \sum_{k=1}^{K} \sum_{i=1}^{I} \sum_{g=1}^{G} \mathbf{O}_{i,g}^k \\ &\text{s.t.}(18b), (18c). \end{aligned} \tag{22}$$

Moreover, since the subproblem of outage probability involves a product of terms, it is challenging to solve directly. To address this, we apply a natural logarithm function to transform the product into a summation, thereby simplifying the problem and facilitating subsequent analysis.

$$\begin{aligned} \mathbf{E}_{i,g}^k = \ln \mathbf{O}_{i,g}^k = \mathrm{Tr}\Bigg\{ &\ln\left(\frac{\lambda_{i,g}^k R_{i,g}^k}{\xi} + L \right) \\ &+ \sum_{z=1}^{i} \frac{\sum_{j=z+1}^{I} p_{j,g}^k |h_{j,g}^k)|^2 + \sigma^2}{-p_{z,g}^k} F_{z,g}^k \Bigg\}, \end{aligned} \tag{23}$$

where $F_{z,g}^k = \frac{(2^{\frac{R_g^g}{B}}-1)\left((1+d_{z,g}^k)^\beta\right)}{2\mu^2}$.

Problem 2 is rewritten as

$$\begin{aligned} &\text{Problem 3:} \max_{p_{i,g}^k} \frac{1}{K} \sum_{k=1}^{K} \sum_{i=1}^{I} \sum_{g=1}^{G} \mathbf{E}_{i,g}^k \\ &\text{s.t.}(18b), (18c). \end{aligned} \tag{24}$$

Since Problem 3 is a non-convex optimization problem with respect to $p_{i,g}^k$, it is challenging to solve directly. Therefore, we adopt the SCA method, whose convergence is proven in [11]. Hence, a lower bound for $R_{i,g}^k$ can be expressed as

$$\begin{aligned} R_{i,g}^k &= B\log_2(1+\mathrm{SINR}_{i,g}^k) \\ &\geq B\left(a_{i,g}^k \log_2(\mathrm{SINR}_{i,g}^k) + b_{i,g}^k\right) = \hat{R}_{i,g}^k, \end{aligned} \tag{25}$$

where the equality holds when

$$a_{i,g}^k = \frac{\mathrm{SINR}_{i,g}^k}{1+\mathrm{SINR}_{i,g}^k}, \tag{26}$$

and

$$b_{i,g}^k = \log_2(1+\mathrm{SINR}_{i,g}^k) - a_{i,g}^k \log_2(\mathrm{SINR}_{i,g}^k). \tag{27}$$

With the fixed approximation coefficients $\mathcal{A}(k) \triangleq a_i^g(k)$ and $\mathcal{B}(k) \triangleq b_{i,g}^k$, the constraint in Problem 3 remains non-concave with respect to $p_{i,g}^k$. Therefore, we introduce the transformation $\bar{p}_{i,g}^k = \ln p_{i,g}^k$.

Here, we decompose the objective function in Problem 3 into two parts, $\mathbf{U}_{i,g}^k$ and $\mathbf{D}_{i,g}^k$, which are expressed as follows:

$$\mathbf{E}_{i,g}^k = \mathrm{Tr}\Big\{\mathbf{U}_{i,g}^k + \mathbf{D}_{i,g}^k\Big\}, \tag{28}$$

$$\mathbf{U}_{i,g}^k = \sum_{z=1}^{i} \frac{\sum_{j=z+1}^{I} e^{\bar{p}_{j,g}^k}|h_{j,g}^k|^2 + \sigma^2}{-e^{\bar{p}_{z,g}^k}} F_{z,g}^k, \tag{29}$$

and

$$\begin{aligned}\mathbf{D}_{i,g}^k &= \ln\left(\frac{\lambda_{i,g}^k R_{i,g}^k}{\xi} + L_{i,g}^k\right)\\ &= \ln\left\{\frac{\lambda_{i,g}^k a_{i,g}^k \bar{p}_{i,g}^k |h_{i,g}^k|^2}{\xi \ln 2} + S_{i,g}^k\right\},\end{aligned} \tag{30}$$

where $\frac{\lambda_{i,g}^k a_{i,g}^k \ln\left(\sum_{j=i+1}^{I} e^{\bar{p}_{j,g}^k}|h_{j,g}^k|^2+\sigma^2\right)+b_{i,g}^k}{-\xi \ln 2} + L_{i,g}^k$ is abbreviated as $S_{i,g}^k$.

The first-order partial derivative of $\mathbf{U}_{i,g}^k$ is given as follows:

$$\frac{\partial \mathbf{U}_{i,g}^k}{\partial(\bar{p}_{i,g}^k)} = \begin{cases} \frac{|h_{i,g}^k|^2}{V_{i,g}^k} F_{i,g}^k, \\ -\sum_{z=1}^{i-1} \delta_{z,g}^k F_{z,g}^k + \frac{|h_{i,g}^k|^2}{V_{i,g}^k} F_{i,g}^k, \\ -\sum_{z=1}^{i-1} \delta_{z,g}^k F_{z,g}^k + \frac{\sigma^2}{e^{\bar{p}_{i,g}^k}} F_{i,g}^k, \end{cases} \tag{31}$$

where we define $V_{i,g}^k = \frac{e^{\bar{p}_{i,g}^k}|h_{i,g}^k|^2}{\sum_{j=i+1}^{I} e^{\bar{p}_{j,g}^k}|h_{j,g}^k|^2+\sigma^2}$ and $\delta_{z,g}^k = \frac{e^{\bar{p}_{i,g}^k}|h_{i,g}^k|^2}{e^{\bar{p}_{z,g}^k}}$. The three cases, listed from top to bottom, correspond to $i = 1$, $1 < i < I$, and $i = I$, respectively.

The second-order partial derivative of $\mathbf{U}_i^g(k)$ is given by

$$\frac{\partial^2 \mathbf{U}_{i,g}^k}{\partial(\bar{p}_{i,g}^k)^2} = \begin{cases} -\frac{|h_{i,g}^k|^2}{V_{i,g}^k} F_{i,g}^k, \\ -\sum_{z=1}^{i-1} \delta_{z,g}^k F_{z,g}^k - \frac{|h_{i,g}^k|^2}{V_{i,g}^k} F_{i,g}^k, \\ -\sum_{z=1}^{i-1} \delta_{z,g}^k F_{z,g}^k - \frac{\sigma^2}{e^{\bar{p}_{i,g}^k}} F_{i,g}^k, \end{cases} \tag{32}$$

where the three constraint conditions for the second-order partial derivatives are the same as those for the first-order derivatives.

The first- and second-order derivatives of $\mathbf{D}_i^g(k)$ are expressed as follows

$$\frac{\partial \mathbf{D}_{i,g}^k}{\partial \bar{p}_{i,g}^k} = \frac{\frac{\lambda_{i,g}^k a_{i,g}^k |h_{i,g}^k|^2}{\xi \ln 2}}{\frac{\lambda_{i,g}^k a_{i,g}^k |h_{i,g}^k|^2}{\xi \ln 2} \bar{p}_{i,g}^k + S_{i,g}^k}, \tag{33}$$

$$\frac{\partial^2 \mathbf{D}_{i,g}^k}{\partial\left(\bar{p}_{i,g}^k\right)^2} = \frac{-\left(\frac{\lambda_{i,g}^k a_{i,g}^k |h_{i,g}^k|^2}{\xi \ln 2}\right)^2}{\left(\frac{\lambda_{i,g}^k a_{i,g}^k |h_{i,g}^k|^2}{\xi \ln 2}\bar{p}_{i,g}^k + S_{i,g}^k\right)^2}. \tag{34}$$

According to Equations (31)–(34), Problem 3 is proven to be a convex optimization problem, and we use the dual decomposition method to solve it. Consequently, the Lagrangian function for Problem 4 is

$$\begin{aligned} &\sum_{i=1}^{I}\sum_{g=1}^{G} \mathcal{L}\left(p_{i,g}^k, \eta_{i,g}^k, \phi_{i,g}^k\right) \\ &= \sum_{i=1}^{I}\sum_{g=1}^{G} \mathbf{E}_{i,g}^k + \sum_{i=1}^{I}\sum_{g=1}^{G} \eta_{i,g}^k\left(p_{\max} - p_{i,g}^k\right) \\ &+ \sum_{i=1}^{I}\sum_{g=1}^{G} \phi_{i,g}^k\left(T_{i,g}^{max} - T_{i,g}^{total,k}\right), \end{aligned} \tag{35}$$

where $\eta_{i,g}^k$ and $\phi_{i,g}^k$ are the Lagrangian multipliers.

According to (35), the dual function $\mathcal{D}\left(\eta_{i,g}^k, \phi_{i,g}^k\right)$ can be expressed as

$$\mathcal{D}\left(\eta_{i,g}^k, \phi_{i,g}^k\right) = \begin{cases} \max\limits_{p_{i,g}^k} \mathcal{L}\left(\bar{p}_{i,g}^k, \eta_{i,g}^k, \phi_{i,g}^k\right) \\ \quad \text{s.t. } \bar{p}_{i,g}^k \geq 0. \end{cases}$$

Hence, the dual problem is

$$\begin{aligned} &\text{Problem 4: } \min \mathcal{D}\left(\eta_{i,g}^k, \phi_{i,g}^k\right) \\ &\quad \text{s.t.} \eta_{i,g}^k \geq 0, \phi_{i,g}^k \geq 0. \end{aligned} \tag{36}$$

The power allocation $\bar{p}_{i,g}^k$ for the fixed values $\eta_{i,g}^k$, $\phi_{i,g}^k$ is calculated via (37) by applying the KKT condition on (35). Hence, we have (see Alogorithm 4)

$$\frac{\partial \mathcal{L}\left(p_{i,g}^k, \eta_{i,g}^k, \phi_{i,g}^k\right)}{\partial \bar{p}_{i,g}^k} = 0. \tag{37}$$

Since the problem involves solving a multi-objective nonlinear equation system where the result is equal to zero, we utilized the lsqnonlin function in MATLAB 2024b to find the optimal solution $\bar{p}_{i,g}^{*,k}$.

For $\eta_{i,g}^k$ and $\phi_{i,g}^k$, we use the gradient descent method to obtain the optimal values:

$$\eta_{i,g}^{j+1,k} = \left[\eta_{i,g}^{j,k} - \Delta\varepsilon_1^{j,k}(p_{\max} - p_{i,g}^k)\right]^+, \tag{38}$$

$$\phi_{i,g}^{j+1,k} = \left[\phi_{i,g}^{j,k} - \Delta\varepsilon_2^{j,k}(T_{i,g}^{max,k} - T_{i,g}^{total,k})\right]^+. \tag{39}$$

where j is an iteration index. The parameters $\Delta\varepsilon_1$ and $\Delta\varepsilon_2$ represent the learning rates for $\eta_{i,g}^j$ and $\phi_{i,g}^j$, respectively.

3.3. Complexity Analysis

The computational complexity of the proposed algorithms was analyzed as follows: The complexity is determined by the number of iterations of the while loop, combined with the nested loops, resulting in a time complexity of $O(t_{\max} \cdot K \cdot I \cdot G)$, where $t_{\max}$ and K represent the maximum number of iterations and the iteration count within the loop, respectively; G is the number of groups; and I is the number of sensors within each group. For Algorithm 3, the complexity is approximately $O(T \cdot N^2 \cdot G)$, where T denotes the number of iterations required to achieve a Nash-stable partition and N represents the total number of sensors. For Algorithms 3, the complexity is $O(K \cdot L \cdot M)$, where L is the number of iterations required for convergence and M is the number of power allocation variables. Thus, the total complexity of the system, combining the complexities of sensor grouping and power allocation processes, is $O(t_{\max} \cdot K \cdot I \cdot G + T \cdot N^2 \cdot G + K \cdot L \cdot M)$. To further verify the practical efficiency of the proposed algorithm, we report the actual computational time of each method under various system scales, as summarized in Table 2.

Algorithm 3 Power Allocation Based on SCA and Dual Decomposition Methods

Input: $\lambda_{i,g}^{\text{iter},k}, \xi, \psi, \text{iter\_max}, \epsilon, D_i^g, R_{i,g}^k, \sigma, i \in \mathcal{I}, g \in \mathcal{G}$.
Initialization: $a_{i,g}^k = b_{i,g}^k = 1$, $\Delta\varepsilon = \Delta\varepsilon_2 = 0.1$, $p_{i,g}^k = p_{\max}$, $j = \text{iter} = 1$, $\mathbf{E}_i^g = \mathbf{E}_{i,g}^k = 0$.

for $k = 1$ to K **do**
  while $\Delta > \epsilon$ **or** iter $\leq$ iter_max **do**
    Update iter $=$ iter $+ 1$
    Update $a_{i,g}^k$ according formula (26)
    Update $b_{i,g}^k$ according formula (27)
    Calculate $p_{i,g}^{*,k} = \exp(\bar{p}_{i,g}^{*,k})$
    repeat
      $j = j + 1$
      Calculate $\bar{p}_{i,g}^k$ according to formula (37)
      Update $\left\{\eta_{i,g}^k\right\}$ according to formula (38)
      Update $\left\{\phi_{i,g}^k\right\}$ according to formula (39)
    until $j \geq$ iter_max
    Update $\Delta = |F(\lambda_{i,g}^{\text{iter},k}) - F(\lambda_{i,g}^{\text{iter-1},k})|$
  end while
  Obtain $\mathbf{E}_{i,g}^k$ according to formula (23)
  Update $\mathbf{E}_i^g = \mathbf{E}_i^g + \mathbf{E}_{i,g}^k$
end for
Output: $\mathbf{E}_i^g = \mathbf{E}_{i,g}^k / K$.

Table 2. Performance comparison under different system scales.

System Scale (N, I, G)	This Work	Random	Suboptimal
$(12, 2, 6)$	1846.6s	1739.0s	1793.1s
$(12, 3, 4)$	1622.3s	1565.7s	1583.0s
$(12, 4, 3)$	1571.0s	1494.9s	1525.4s

4. Performance Evalution

In this section, we evaluate the performance of the resource allocation algorithm for minimizing outage risk under the joint optimization of remote state estimation and transmission delay. The experiments were conducted using MATLAB 2024b. The simulation setup positioned the base station at coordinates $(0\,\text{m}, 0\,\text{m})$, with sensors randomly distributed within a $100\,\text{m} \times 100\,\text{m}$ area. This area size was chosen to represent a typical industrial Internet of Things (IIoT) scenario, where a large number of sensors operate in a compact space with short-range wireless communication and strict latency and reliability requirements. We simulated 1000 Rayleigh fading channel realizations and used the sampled values to calculate channel gains. Each sensor was assigned a bandwidth of 1 KHz for data transmission. To reflect a dynamic and time-varying wireless environment, we further assumed that sensor positions were updated at each time slot, simulating node mobility or deployment variations commonly encountered in practical wireless sensor networks. Specific parameters are summarized in Table 3. The system parameters A_i^g, C_i^g, Q_i^g, and R_i^g were sampled from the range $[0.5, 1.5]$ to reflect moderate system dynamics and noise levels while ensuring numerical stability for Kalman filtering. The optimization-related parameters were set as $\zeta = 0.4$, $\psi = 1$, and $\mu = 1$, representing a balanced trade-off between communication delay and estimation accuracy, as well as a standard Rayleigh fading environment. These parameter settings were consistent with prior works and supported the generality and reliability of the simulation.

Table 3. Summary of the simluation parameters.

Parameters	Values
Simulation area size	100 m × 100 m
Base station antenna location	(0, 0) m
CPU cycles required to process task I_i^g	1000 Megacycles
Computational capability of sensor F^s	1 GHz
Computational capability of base station F^b	10 GHz
System bandwith B	1 KHz
Minimum transmission rate $\hat{R}$	100 bps
Maximum power $p_{\max}$	20 dBm
Maximum tolerable transmission delay $T_{i,g}^{max}$	0.1 s
Size of received input data D_i^g	80 B
Noise power density σ	−170 dBm/Hz
State transition matrix A_i^g	[0.5, 1.5]
Measurement matrix C_i^g	[0.5, 1.5]
Process noise covariance Q_i^g	[0.5, 1.5]
Measurement noise covariance R_i^g	[0.5, 1.5]
Weighting factor ξ	0.4
Scaling factor ψ	1
Variance of the normal distribution μ	1

For the sensor grouping algorithm, we evaluated two alternative strategies and compared them with the proposed heuristic algorithm, which optimized sensor grouping under fixed power constraints to minimize the state estimation outage risk. The two strategies were random sensor grouping [28] and a suboptimal grouping approach [29]. The suboptimal strategy operated as follows: given N sensors, they were first sorted in descending order based on their channel gains. The sorted sensors were then evenly divided into M blocks. Sensors from the same position in each block formed a group (e.g., the first sensor in each block formed the first group, the second sensor formed the second group, etc.) until all groups were created. For the power allocation algorithm, we compared two approaches: random power allocation within the allowable range [30] and assigning the maximum power value to all sensors [15]. Comparing the algorithm with the four aforementioned algorithms helped evaluate the performance, computational complexity, and practical feasibility of different strategies in the uplink NOMA system, thereby optimizing resource allocation.

Figure 2 presents the sum of the expected function in Problem 4 with nine sensors and three channels. The simulation parameters were set as $\xi = 0.5$, $\hat{R} = 100$ bps, $B = 1$ KHZ, and $p_{\max} = 20$ dBm. We compared different sensor grouping strategies, including the proposed coalition game-based intelligent grouping algorithm, the random grouping algorithm, and the suboptimal grouping algorithm. As shown, across all time slots k, the coalition game-based algorithm consistently outperformed the other two in solving the sum of expectations in Problem 1. Each point was obtained from 500 Monte Carlo simulations per algorithm per time slot.

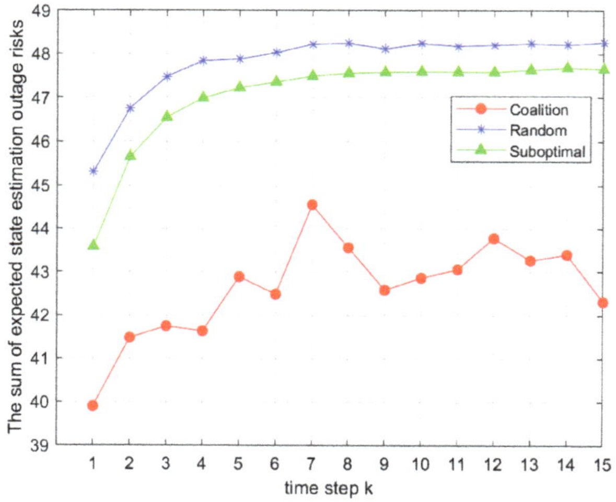

Figure 2. Comparison of the total outage risks under different sensor grouping strategies.

Figure 3 compares different power allocation methods under the same coalition game-based grouping strategy. Simulation settings and Monte Carlo parameters were consistent with those in Figure 2. The joint Dinkelbach algorithm, the SCA algorithm, and the KKT-based power allocation were evaluated against random and full power allocation. The proposed joint allocation methods clearly outperformed the baselines in solving the sum of expectations in Problem 1.

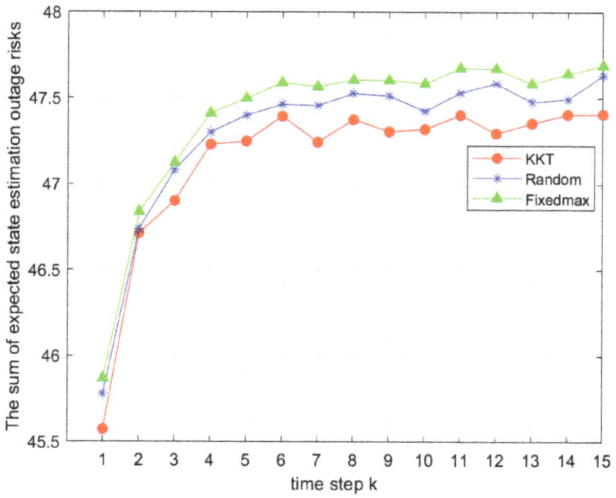

Figure 3. Comparison of the total outage risks under different power allocation strategies.

Figure 4 illustrates the impact of different minimum transmission rates on the total value of the objective function in Problem 1. The minimum rates were set to 100 bps, 500 bps, and 1000 bps, while other parameters remained unchanged. It was observed that the total objective value increased significantly with $\hat{R}_{i,g}^{k}$, which aligned with the theoretical expectations based on Equation (20).

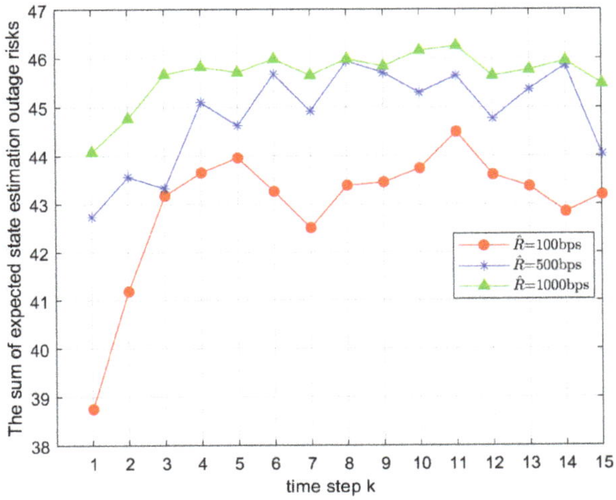

Figure 4. Comparison of the total outage risks under different minimum transmission rates $\hat{R}$.

Figure 5 examines the effect of bandwidth B on $\mathbf{J}_{i,g}^{k}$. Bandwidth values were set to 1 KHz, 10 KHz, and 100 KHz. The curves show that as B increased, $\mathbf{J}_{i,g}^{k}$ grew accordingly. Limited bandwidth constrained communication performance, while a larger B improved transmission rates and system efficiency, validating the model's sensitivity to bandwidth variations.

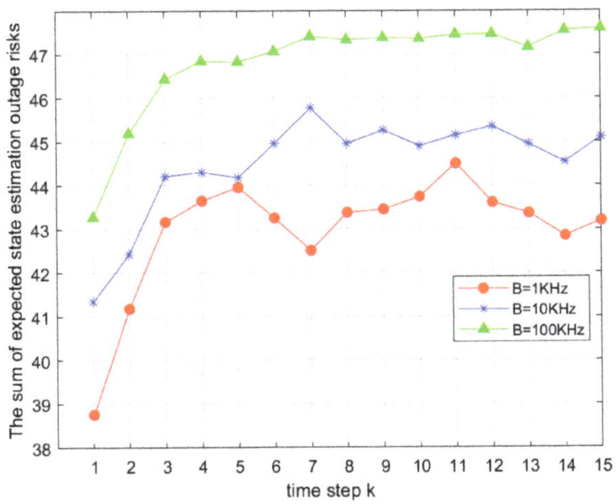

Figure 5. Comparison of the total outage risks under different bandwidths B.

In Figure 6, we show the impact of the weighting parameter ξ on the total value $\mathbf{J}_{i,g}^{k}$ of the objective function in Problem 1. Specifically, we varied the parameter ξ with values $\xi = 0.2$, $\xi = 0.5$, and $\xi = 0.8$. The results indicate a clear trend: as the proportion allocated to ξ increased, the total value $\mathbf{J}_{i,g}^{k}$ also rose. This suggests that in the joint optimization of remote state estimation using KF and the offloading time consumption, the latter has a more pronounced influence on the objective function. Given that our optimization problem aimed to minimize the objective function, this result highlights a critical trade-off. To achieve optimal system performance, it is advisable to assign a relatively smaller weight to offloading time consumption while allocating a greater proportion to the remote state estimation component. This allocation strategy aligns with the objective of minimizing the overall efficiency and benefit of the system.

In Figure 6, the impact of the weighting parameter ξ on $\mathbf{J}_{i,g}^{k}$ is shown. We considered 0.2, 0.5, and 0.8. The results show that as ξ increased, the total objective value rose. This indicates that offloading time consumption has a greater influence on the objective, suggesting that assigning a smaller weight to offloading time can better optimize system performance.

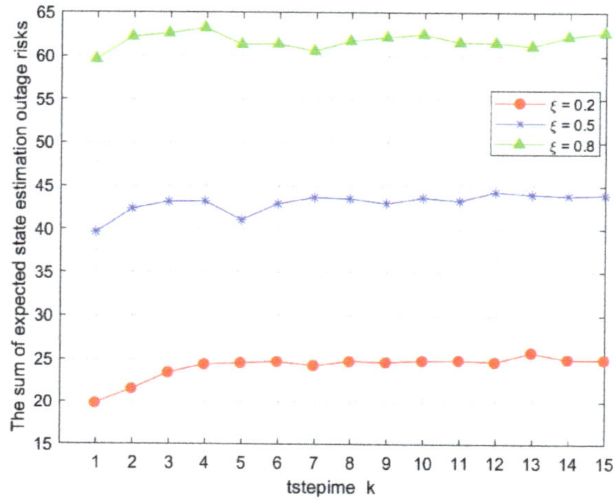

Figure 6. Comparison of total outage risks under different association factors ξ.

Figure 7 analyzes the impact of maximum transmission power on $\mathbf{J}_{i,g}^{k}$. As $p_{\max}$ increased, the total objective value decreased. This was because higher transmission power strengthened the transmitted signal, thereby improving transmission reliability according to the Shannon–Hartley theorem. However, the resulting increase in power consumption must be considered in practical applications to balance performance and energy efficiency.

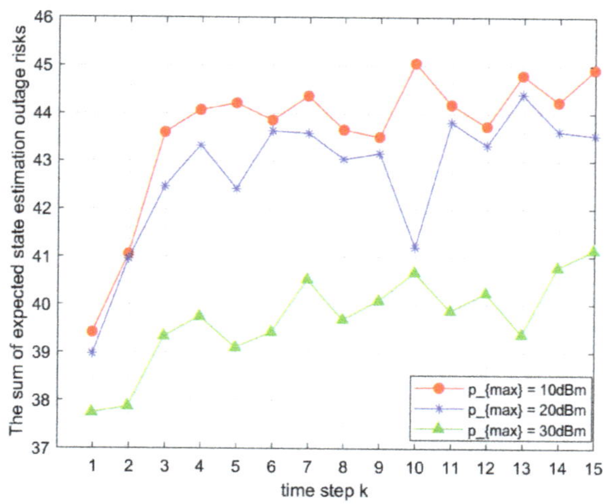

Figure 7. Comparison of total outage risks under different maximum powers $p_{\max}$.

In Figure 8, we analyze how different sensor allocations within each channel group affected $\mathbf{J}_{i,g}^{k}$ with a fixed total of 12 sensors. Group sizes of two, three, and four sensors were compared. The results show that increasing the number of sensors per group improved system performance by reducing the total objective value. However, it significantly increased computational complexity, with computation time for four sensors per group being approximately 10 times higher than for two or three sensors. Thus, a trade-off between performance and computational cost must be carefully considered.

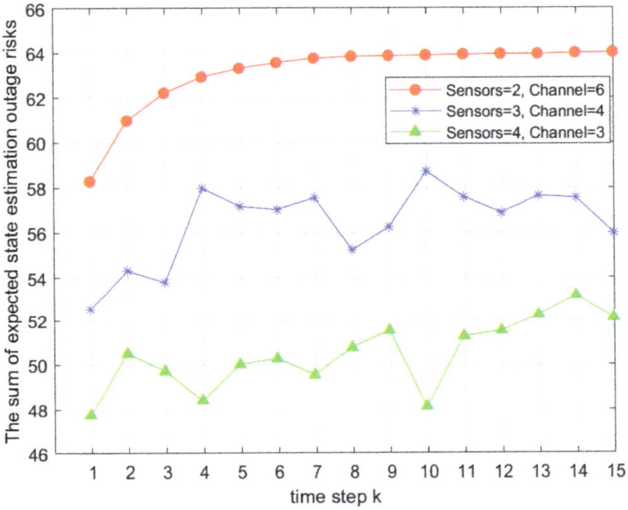

Figure 8. Comparison of total outage risks under different sensor counts within the same channel.

5. Conclusions

This paper investigated the problem of minimizing outage risk under joint optimization of remote state estimation and transmission delay. By formulating the problem as a MINLP problem, it was further divided into an intelligent sensor grouping algorithm based on coalition game theory and a joint power allocation algorithm leveraging the Dinkelbach algorithm, SCA, and dual decomposition methods. Simulation results demonstrate that the proposed algorithm outperforms traditional sensor grouping and power allocation algorithms in terms of performance, effectively reducing the outage risk of sensors during transmission. Future research can further extend to multi-sensor and multi-base station scenarios in downlink NOMA systems, exploring outage risk assessment methods in more complex systems.

Author Contributions: Conceptualization, R.L.; methodology, R.L. and L.X.; software, R.L.; validation, L.X.; formal analysis, R.L.; investigation, R.L.; resources, R.L.; data curation, R.L.; writing—original draft preparation, R.L.; writing—review and editing, R.L. and L.X.; visualization, R.L.; supervision, R.L.; project administration, L.X. All authors have read and agreed to the published version of the manuscript.

Funding: This research received no external funding.

Institutional Review Board Statement: Not applicable.

Informed Consent Statement: Not applicable.

Data Availability Statement: The data that support the findings of this study are available from the corresponding author upon reasonable request.

Conflicts of Interest: The authors declare no conflict of interest.

References

1. Salama, R.; Al-Turjman, F.; Bordoloi, D.; Yadav, S.P. Wireless Sensor Networks and Green Networking for 6G communication—An Overview. In Proceedings of the 2023 International Conference on Computational Intelligence, Communication Technology and Networking (CICTN), Ghaziabad, India, 20–21 April 2023; pp. 830–834. [CrossRef]
2. Khan, W.U.; Jameel, F.; Jamshed, M.A.; Pervaiz, H.; Khan, S.; Liu, J. Efficient power allocation for NOMA-enabled IoT networks in 6G era. *Phys. Commun.* **2020**, *39*, 101043. [CrossRef]
3. Sarma, S.S.; Sachan, A.; Hazra, R.; Talukdar, F.A.; Mukherjee, A.; Chatterjee, P.; Al-Numay, W. D2D Communication in a 5G mm-Wave Cellular Network for Wireless Sensor Networks. *IEEE Sens. J.* **2024**, *24*, 5512–5521. [CrossRef]
4. Rathod, T.; Gupta, R.; Nehra, A.; Kumar Jadav, N.; Nehra, A. AI and Coalition Game Interplay for Efficient Resource Allocation in D2D Communication. In Proceedings of the GLOBECOM 2023—2023 IEEE Global Communications Conference, Lumpur, Malaysia, 4–8 December 2023; pp. 3058–3063. [CrossRef]
5. Hu, H.; Ma, C.; Tang, B. Channel Allocation Scheme for Ultra-Dense Femtocell Networks: Based on Coalition Formation Game and Matching Game. In Proceedings of the 2019 IEEE MTT-S International Wireless Symposium (IWS), Guangzhou, China, 19–22 May 2019; pp. 1–3. [CrossRef]
6. Chen, W.; Zhao, S.; Zhang, R.; Yang, L. Generalized User Grouping in NOMA Based on Overlapping Coalition Formation Game. *IEEE J. Sel. Areas Commun.* **2021**, *39*, 969–981. [CrossRef]
7. Liu, G.; Wang, R.; Zhang, H.; Kang, W.; Tsiftsis, T.A.; Leung, V.C.M. Super-Modular Game-Based User Scheduling and Power Allocation for Energy-Efficient NOMA Network. *IEEE Trans. Wirel. Commun.* **2018**, *17*, 3877–3888. [CrossRef]
8. Liu, B.; Su, Z.; Xu, Q. Game theoretical secure wireless communication for UAV-assisted vehicular Internet of Things. *China Commun.* **2021**, *18*, 147–157. [CrossRef]
9. Wang, D.; Huang, J.; He, M.; Huang, C. Spectrum Transaction Games for UAV Assisted Communications. *IEEE Wirel. Commun. Lett.* **2022**, *11*, 1216–1219. [CrossRef]
10. Yang, Q.; Zhang, Q.; Peng, Y. Cluster-Based Strategy for Maximizing the Sum-Rate of a Distributed Reconfigurable Intelligent Surface (RIS)-Assisted Coordinated Multi-Point Non-Orthogonal Multiple-Access (CoMP-NOMA) System. *Sensors* **2024**, *24*, 3644. [CrossRef] [PubMed]
11. Xu, L.; Zhou, Y.; Wang, P.; Liu, W. Max-Min Resource Allocation for Video Transmission in NOMA-Based Cognitive Wireless Networks. *IEEE Trans. Commun.* **2018**, *66*, 5804–5813. [CrossRef]
12. Azarhava, H.; Musevi Niya, J. Energy Efficient Resource Allocation in Wireless Energy Harvesting Sensor Networks. *IEEE Wirel. Commun. Lett.* **2020**, *9*, 1000–1003. [CrossRef]
13. Liu, M.; Wu, Q.; Wang, Z.; Zhao, B.; Zhang, L.; Li, J.; Zhu, X. Optimal power allocation strategy for scaled hydrogen storage system considering power-efficiency coupling relationship. In Proceedings of the 2023 IEEE Sustainable Power and Energy Conference (iSPEC), Chongqing, China, 28–30 November 2023; pp. 1–5. [CrossRef]
14. Zamani, M.R.; Eslami, M.; Khorramizadeh, M.; Zamani, H.; Ding, Z. Optimizing Weighted-Sum Energy Efficiency in Downlink and Uplink NOMA Systems. *IEEE Trans. Veh. Technol.* **2020**, *69*, 11112–11127. [CrossRef]
15. Sultan, R.; Song, L.; Seddik, K.G.; Han, Z. Joint Resource Management With Distributed Uplink Power Control in Full-Duplex OFDMA Networks. *IEEE Trans. Veh. Technol.* **2020**, *69*, 2850–2863. [CrossRef]
16. Zhang, Q.; An, K.; Yan, X.; Xi, H.; Wang, Y. User Pairing for Delay-Limited NOMA-Based Satellite Networks with Deep Reinforcement Learning. *Sensors* **2023**, *23*, 7062. [CrossRef] [PubMed]
17. Feng, J.; Yu, F.R.; Pei, Q.; Du, J.; Zhu, L. Joint Optimization of Radio and Computational Resources Allocation in Blockchain-Enabled Mobile Edge Computing Systems. *IEEE Trans. Wirel. Commun.* **2020**, *19*, 4321–4334. [CrossRef]
18. Dixit, S.; Shukla, V.; Misra, M.K.; Jimenez, J.M.; Lloret, J. Progressive Pattern Interleaver with Multi-Carrier Modulation Schemes and Iterative Multi-User Detection in IoT 6G Environments with Multipath Channels. *Sensors* **2024**, *24*, 3648. [CrossRef] [PubMed]
19. Tabee Miandoab, F.; Fazel, M.S.; Mahdavi, M. Outage Analysis of Multiuser MIMO-NOMA Transmissions in Uplink Full-Duplex Cooperative System. *IEEE Wirel. Commun. Lett.* **2022**, *11*, 2076–2079. [CrossRef]
20. Lu, H.; Xie, X.; Shi, Z.; Cai, J. Outage Performance of CDF-Based Scheduling in Downlink and Uplink NOMA Systems. *IEEE Trans. Veh. Technol.* **2020**, *69*, 14945–14959. [CrossRef]
21. Lu, H.; Xie, X.; Shi, Z.; Kadoch, M.; Cheriet, M.; Cai, J. Outage Probability of CDF-Based Scheduling for Uplink NOMA with Practical SIC Considerations. In Proceedings of the 2020 International Wireless Communications and Mobile Computing (IWCMC), Limassol, Cyprus, 15–19 June 2020; pp. 1031–1036. [CrossRef]
22. Liu, H.; Tsiftsis, T.A.; Kim, K.J.; Kwak, K.S.; Poor, H.V. Rate Splitting for Uplink NOMA With Enhanced Fairness and Outage Performance. *IEEE Trans. Wirel. Commun.* **2020**, *19*, 4657–4670. [CrossRef]

23. Yadav, P.; Jain, S. Outage Probability Analysis of IRS-Assisted NOMA System over Rician Fading Channel. In Proceedings of the 2024 15th International Conference on Computing Communication and Networking Technologies (ICCCNT), Kamand, India, 24–28 June 2024; pp. 1–6. [CrossRef]
24. Basha, S.T.; Nageena Parveen, S.; Bathini, A.; Gade, S.S.; Kothapelly, V. Outage Analysis of Downlink Cooperative SWIPT -NOMA system with optimum energy harvesting and imperfect CSI. In Proceedings of the 2024 15th International Conference on Computing Communication and Networking Technologies (ICCCNT), Kamand, India, 24–28 June 2024; pp. 1–6. [CrossRef]
25. Trankatwar, S.; Wali, P. Optimal Power Allocation for Downlink NOMA Heterogeneous Networks to Improve Sum Rate and Outage Probability. In Proceedings of the 2022 IEEE India Council International Subsections Conference (INDISCON), Bhubaneswar, India, 15–17 July 2022; pp. 1–6. [CrossRef]
26. Li, Y.; Mehr, A.S.; Chen, T. Multi-sensor transmission power control for remote estimation through a SINR-based communication channel. *Automatica* **2019**, *101*, 78–86. [CrossRef]
27. Pang, G.; Liu, W.; Li, Y.; Vucetic, B. Deep Reinforcement Learning for Radio Resource Allocation in NOMA-based Remote State Estimation. In Proceedings of the GLOBECOM 2022—2022 IEEE Global Communications Conference, Rio de Janeiro, Brazil, 4–8 December 2022; pp. 3059–3064. [CrossRef]
28. Ahmad, A.W.; Mehmood Bahadar, N. Exploiting Heterogeneous Networks model for Cluster Formation and Power Allocation in Uplink NOMA. In Proceedings of the 2019 21st International Conference on Advanced Communication Technology (ICACT), PyeongChang, Republic of Korea, 17–20 February 2019; pp. 129–133. [CrossRef]
29. Mohsenivatani, M.; Liu, Y.; Derakhshani, M.; Parsaeefard, S.; Lambotharan, S. Completion-Time-Driven Scheduling for Uplink NOMA-Enabled Wireless Networks. *IEEE Commun. Lett.* **2020**, *24*, 1775–1779. [CrossRef]
30. Duan, Z.; Yang, X.; Xu, Q.; Wang, L. Covert Communication in Uplink NOMA Systems Against a Two-Phase Detector. In Proceedings of the GLOBECOM 2022—2022 IEEE Global Communications Conference, Rio de Janeiro, Brazil, 4–8 December 2022; pp. 5516–5521. [CrossRef]

sensors

MDPI

Communication

Multiple-Cumulant-Matrices-Based Method for Exact NF Polarization Localization with COLD Array

Jiefeng Zheng [1,†], Haifen Meng [1], Zhuang Luo [1,*,†], Huayue Wu [2], Weiyue Liu [1] and Hua Chen [1]

[1] Faculty of Electrical Engineering and Computer Science, Ningbo University, Ningbo 315211, China; zhengjf7663@163.com (J.Z.); 19807857024@163.com (H.M.); liuweiyue@nbu.edu.cn (W.L.); dkchenhua0714@hotmail.com (H.C.)

[2] School of Materials Science and Chemical Engineering, Ningbo University, Ningbo 315211, China; 236003061@nbu.edu.cn

* Correspondence: 19958357701@163.com

† These authors contributed equally to this work.

Abstract: As a key technology for the fifth-generation of mobile communications, massive MIMO systems enable massive user access via large-scale arrays. However, their dense deployment extends the near-field (NF) region, introducing new localization complexities. Based on an exact spherical wavefront model, this paper proposes a multiple-cumulant-matrices-based method for NF source localization using a Co-centered Orthogonal Loop and Dipole (COLD) array. Firstly, following the physical numbering of array elements, we can construct multiple polarization cumulant matrices, which can then be cascaded into a long matrix. Next, the signal subspace can be obtained through eigen-decomposition of this long matrix, from which the horizontal and vertical components can be further separated. By applying ESPRIT, joint angle, range, and polarization parameters can be estimated. In addition, the asymptotic variances for joint spatial and polarization parameters are analyzed. Compared with existing NF polarization algorithms, the proposed method exhibits better parameter estimation and is consistent with a theoretical asymptotic performance.

Keywords: spherical wavefront; fourth order cumulant; ESPRIT; asymptotic variance

1. Introduction

In array signal processing, source localization is an important problem, with widespread applications in wireless communication and radar systems [1–7]. In practical scenarios, such as multiple-input multiple-output (MIMO) systems equipped with a large antenna array, the target source crosses the Fresnel boundary, entering the NF region. When the sources move from the far-field (FF) region, the planar wavefront assumption is not available, requiring characterization by a spherical wavefront that is bivariate in both the Direction-of-Arrival (DOA) and range parameters, and the propagation attenuation induces range-dependent amplitude variations among sensors, thereby rendering FF localization methods ineffective.

Nowadays, numerous algorithms for FF localization have been successfully generalized for NF applications, such as the maximum-likelihood (ML) method [8], second-order statistics (SOS) methods [9–12] and fourth-order cumulant (FOC) methods [13–15]. Based on the wavefront approximation principle, i.e., Fresnel approximation, these methods suffer from amplitude and phase loss compared with practical wireless environments [16–18]. For amplitude attenuation, several works have addressed this issue. In [19], a delay-correlation SOS-based method was proposed considering amplitude attenuation; however, to satisfy the symmetric property of the array, this method utilizes an approximation phase item. Recently, based on an exact spherical wavefront model, Ref. [20] proposed a delay FOC-based method, which can achieve underdetermined estimation by using a cumulant matrix and vectorization operation, but it requires 3-D spectral search, suffering from an extremely high complexity. In [21], a mixed FF and NF localization method was proposed using coarse and fine estimation from the amplitude and phase item, and this method is closed-from, but suffers from aperture loss.

Although all the abovementioned algorithms were designed for effective spatial parameter estimation, they neglect polarization information, which is a fundamental feature of electromagnetic sources. With the introduction of polarization-sensitive arrays, both spatial and polarization parameter localization are possible. In [22], a generalized ESPRIT method was performed for NF polarization source localization based on a cross-dipole array, which can decouple spatial and polarization parameters, but it can only estimate spatial parameters. In [23], a rank-reduction-based method was proposed, but it still required a 2-D spectral search to estimate polarization parameters. The aforementioned methods [8–11,13,15,22,23] employed Fresnel approximation. Then, a COLD array-based exact NF polarization localization method was proposed in [24], which can estimate the polarization rotational matrix, but estimates spatial parameters from range-dependent amplitude attenuation, resulting in a lower estimation accuracy.

To address the issue of exact NF polarization source localization, by employing a COLD array, this paper proposes a multiple-cumulant-based DOA estimation method. Cumulant matrices related to various array elements are calculated, which can then be combined in a long matrix, avoiding the aperture loss. By using ESPRIT, the proposed method can achieve spatial and polarization decoupling and joint estimation of four-dimensional (4-D) parameters. Moreover, joint spatial and polarization parameter performance evaluation was performed to demonstrate the algorithm's validity.This paper represents an original contribution in three aspects:

(1) Unlike conventional NF polarization localization methods, the proposed algorithm completely avoids systematic errors induced by model approximations.

(2) Multiple cumulant matrices are constructed and cascaded into a long matrix, so the manifold matrix for joint parameters can be estimated without aperture loss.

(3) We derive an analytical expression for the asymptotic variance under an exact NF polarization condition.

2. Signal Model

As depicted in Figure 1, assume that there is a linear array deployed along the x-axis consisting of $2L+1$ COLD antennas, where the center antenna, i.e., the 0-th antenna, serves as amplitude-phase reference. Consider K narrowband completely polarized signals impinging on the array from $(\theta_1, r_1), \cdots, (\theta_k, r_k), \cdots (\theta_K, r_K)$ with corresponding polarization parameters $(\gamma_1, \eta_1), \cdots, (\gamma_k, \eta_k), \cdots, (\gamma_K, \eta_K)$, where θ_k, r_k, γ_k and η_k represent the angle, range, polarization angle, and polarization phase difference of the k-th signal $s_k(t)$, respectively.

Based on spherical wavefront geometry, the range between the k-th source and l-th sensor is given by

$$r_{l,k} = \sqrt{r_k^2 + d_l^2 - 2r_k d_l \sin\theta_k}, \tag{1}$$

where d_l represents the distance between the l-th COLD and reference point, $l = -L, ..., 0, ..., L$. Therefore, the exact spatial amplitude-phase corresponding to between k-th source and l-th sensor be expressed as

$$b_{l,k} = \frac{r_k}{r_{l,k}} e^{j\frac{2\pi}{\lambda}(r_k - r_{l,k})} = \rho_{l,k} e^{j\tau_{l,k}}, \tag{2}$$

where $\rho_{l,k}$ stands for the amplitude attenuation, $\tau_{l,k} = \frac{2\pi}{\lambda}(r_k - r_{l,k})$ stands for the exact phase shift caused by propagation delay, and then we can obtain the spatial steering vector of the k-th source, which can be expressed as

$$\mathbf{b}(\theta_k, r_k) = [b_{-L,k}, ..., b_{0,k}, ..., b_{L,k}]^T. \tag{3}$$

Additionally, for the COLD array, the polarization steering vector of the k-th source can be written as

$$\mathbf{g}(\gamma_k, \eta_k) = \begin{bmatrix} \sin \gamma_k e^{j\eta_k} \\ \cos \gamma_k \end{bmatrix}. \tag{4}$$

Hence, we can obtain a spatial and polarization joint steering vector as

$$\mathbf{a}(\theta_k, r_k, \gamma_k, \eta_k) = \mathbf{g}(\gamma_k, \eta_k) \otimes \mathbf{b}(\theta_k, r_k), \tag{5}$$

where $\otimes$ represents the Kronecker product, $\mathbf{a}(\theta_k, r_k, \gamma_k, \eta_k)$ is the $2(2L+1) \times 1$ vector. At time t, the signal data vector of the COLD array can be expressed as the following $2(2L+1) \times 1$ vector:

$$\begin{aligned} \mathbf{x}(t) &= \sum_{k=1}^{K} \mathbf{g}(\gamma_k, \eta_k) \otimes \mathbf{b}(\theta_k, r_k) s_k(t) + \mathbf{n}(t) \\ &= (\mathbf{G} \odot \mathbf{B})\mathbf{s}(t) + \mathbf{n}(t) \\ &= \mathbf{A}\mathbf{s}(t) + \mathbf{n}(t) \end{aligned}, \tag{6}$$

where $\odot$ represents the Khatri–Rao product, $\mathbf{A} = [\mathbf{a}(\theta_1, r_1, \gamma_1, \eta_1), ..., \mathbf{a}(\theta_K, r_K, \gamma_K, \eta_K)]$ signifies the $2(2L+1) \times K$ manifold matrix, $\mathbf{B} = [\mathbf{b}(\theta_1, r_1), ..., \mathbf{b}(\theta_K, r_K)]$ stands for spatial manifold matrix, $\mathbf{G} = [\mathbf{g}(\gamma_1, \eta_1), ..., \mathbf{g}(\gamma_K, \eta_K)]$ stands for the $2 \times K$ polarization manifold matrix, $\mathbf{s}(t) = [s_1(t), ..., s_K(t)]^T$ represents the $K \times 1$ signal vector, and $\mathbf{n}(t)$ represents the $2(2L+1) \times 1$ additive noise vector of COLD array with zero mean.

From (5), the signal data vector can be divided into horizontal and vertical components, expressed as

$$\mathbf{x}(t) = \begin{bmatrix} \mathbf{x}_1(t) \\ \mathbf{x}_2(t) \end{bmatrix} = \begin{bmatrix} \mathbf{BP}_1 \\ \mathbf{BP}_2 \end{bmatrix} \mathbf{s}(t) + \mathbf{n}(t), \tag{7}$$

where $\mathbf{x}_1(t)$ is composed of the first $2L+1$ rows of $\mathbf{x}(t)$, $\mathbf{x}_1(t) = [x_{1,-L}(t), ..., x_{1,L}(t)]^T$, and $\mathbf{x}_2(t)$ containing the last $2L+1$ rows, $\mathbf{x}_2(t) = [x_{2,-L}(t), ..., x_{2,L}(t)]^T$. $\mathbf{P}_1$ and $\mathbf{P}_2$ are diagonal matrices, whose k-th diagonal elements are $\sin \gamma_k e^{j\eta_k}$ and $\cos \gamma_k$, respectively. Hence, the steering vector $\mathbf{a}(\theta_k, r_k, \gamma_k, \eta_k)$ can be rewritten as two components, i.e., $\mathbf{a}(\theta_k, r_k, \gamma_k, \eta_k) = [a_{1,-L,k}, ..., a_{1,L,k}, a_{2,-L,k}, ..., a_{2,L,k}]^T$, where $a_{1,l,k} = \sin \gamma_k e^{j\eta_k} b_{l,k}$ and $a_{2,l,k} = \cos \gamma_k b_{l,k}$.

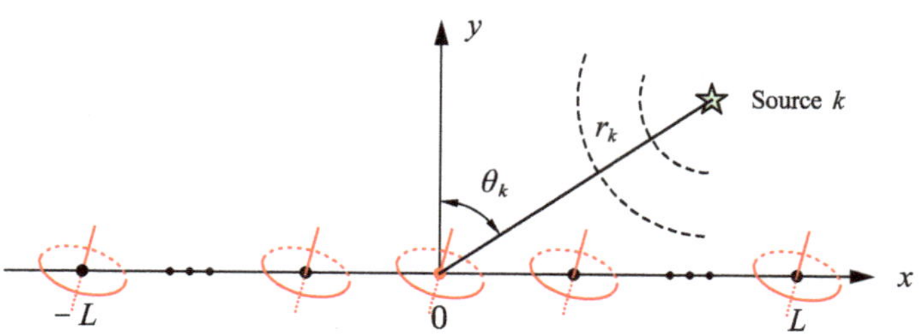

Figure 1. COLD array configuration for exact NF source localization.

3. The Proposed Method

According to the definition in [25], the FOC of $x_p(t), x_q(t), x_i(t), x_j(t)$ is given by

$$C\left(x_p, x_q^*, x_i, x_j^*\right) = E\{x_p x_q^* x_i x_j^*\} \\ - E\{x_p x_q^*\}E\{x_i x_j^*\} \\ - E\{x_p x_i\}E\{x_q^* x_j^*\} \\ - E\{x_p x_j^*\}E\{x_q^* x_i\} \tag{8}$$

3.1. Spatial Manifold Matrix Estimation

Using the signal data vector, we can construct the FOC matrix of the l-th COLD antenna $\mathbf{C}_l$ as follows [26]

$$\begin{aligned} \mathbf{C}_l &= \sum_{i=1}^{2} \mathrm{cum}\left(x_{i,l}(t), x_{i,0}^*(t), \mathbf{x}(t), \mathbf{x}^H(t)\right) \\ &= \sum_{i=1}^{2} \left\{ \sum_{k=1}^{K} a_{i,l,k} a_{i,0,k} \mathbf{a}_k \mathbf{a}_k^H \times \mathrm{cum}(s_k(t), s_k^*(t), s_k(t), s_k^*(t)) \right\}, \\ &= \sum_{k=1}^{K} b_{l,k} \mathbf{a}_k \mathbf{a}_k^H c_{4s_k} \end{aligned} \tag{9}$$

where $(\cdot)^*$ represents complex conjugate, $(\cdot)^H$ represents conjugate transpose, and the FOC matrix $\mathbf{C}_l$ can be expressed in compact form as $\mathbf{C}_l = \mathbf{A}\mathbf{\Phi}_l \mathbf{C}_{4s} \mathbf{A}^H$ with dimensions of $2(2L+1) \times 2(2L+1)$, $c_{4s_k} = \mathrm{cum}(s_k(t), s_k^*(t), s_k(t), s_k^*(t))$ stands for the FOC of k-th signal, $\mathbf{\Phi}_l$ is diagonal matrix, whose diagonal elements are related to the amplitude-phase corresponding of l-th COLD array, $\mathbf{\Phi}_l = \mathrm{diag}\left[\rho_{l,1} e^{j\tau_{l,1}}, \ldots, \rho_{l,K} e^{j\tau_{l,K}}\right]$, $\mathbf{C}_{4s} = \mathrm{diag}\left[c_{4s_1}, \ldots, c_{4s_K}\right]$.

From (9), for the complete COLD array, we can construct $2L+1$ similar cumulant matrices by varying the sensor index. Subsequently, a long matrix Ξ can be obtained by cascading these cumulant matrices

$$\Xi = \begin{bmatrix} \mathbf{C}_{-L} \\ \vdots \\ \mathbf{C}_0 \\ \vdots \\ \mathbf{C}_L \end{bmatrix} = \begin{bmatrix} \mathbf{A}\mathbf{\Phi}_{-L}\mathbf{C}_{4s}\mathbf{A}^H \\ \vdots \\ \mathbf{A}\mathbf{\Phi}_0\mathbf{C}_{4s}\mathbf{A}^H \\ \vdots \\ \mathbf{A}\mathbf{\Phi}_L\mathbf{C}_{4s}\mathbf{A}^H \end{bmatrix}, \tag{10}$$

where $\mathbf{\Phi}_0$ is the identity matrix. The covariance matrix of this joint long matrix is given by

$$\mathbf{R} = E\{\Xi\Xi^H\}. \tag{11}$$

The eigen-decomposition of the covariance matrix $\mathbf{R}$ yields

$$\mathbf{R} = \mathbf{U}\mathbf{\Sigma}\mathbf{U}^H = \mathbf{U}_\mathrm{S}\mathbf{\Sigma}_\mathrm{S}\mathbf{U}_\mathrm{S}^H + \mathbf{U}_\mathrm{N}\mathbf{\Sigma}_\mathrm{N}\mathbf{U}_\mathrm{N}^H, \tag{12}$$

where $\mathbf{U}$ contains the eigenvectors of $\mathbf{R}$, $\mathbf{\Sigma}$ is the diagonal matrix, the signal subspace $\mathbf{U}_\mathrm{S}$ corresponds to the K largest eigenvalues with the dimensions $2(2L+1)^2 \times K$, while the noise subspace $\mathbf{U}_\mathrm{N}$ corresponds to the $2(2L+1)^2 - K$ smallest eigenvalues.

The signal subspace $\mathbf{U}_S$ comprises the subspace associated with the cumulant matrices of all array elements, which can be partitioned into $2L+1$ block sub-matrices as follows:

$$\mathbf{U}_S = \left[\mathbf{U}_{S_{-L}}^T, \cdots, \mathbf{U}_{S_0}^T, \cdots, \mathbf{U}_{S_L}^T\right]^T, \tag{13}$$

where $\mathbf{U}_{S_0}$, formed by the elements from rows $2(2L+1)L+1$ to $2(2L+1)L+1$ of $\mathbf{U}_S$, represents the subspace corresponding to the cumulant matrix of the reference. Due to the rotation-invariant property between the reference cumulant matrix and other sensors' matrices, there exists a full-rank matrix $\mathbf{T}$ satisfying

$$\mathbf{U}_S = \begin{bmatrix} \mathbf{A\Phi}_{-L} \\ \vdots \\ \mathbf{A\Phi}_0 \\ \vdots \\ \mathbf{A\Phi}_L \end{bmatrix} \mathbf{T}. \tag{14}$$

From (14), between $\mathbf{U}_{S_l}$ and $\mathbf{U}_{S_0}$, we can have the following relationship as

$$\mathbf{U}_{S_l} = \mathbf{U}_{S_0}\mathbf{T}^{-1}\mathbf{\Phi}_l\mathbf{T} = \mathbf{U}_{S_0}\mathbf{\Psi}_l, \tag{15}$$

where $\mathbf{\Psi}_l = \mathbf{T}^{-1}\mathbf{\Phi}_l\mathbf{T}$, $\mathbf{\Phi}_l$ is diagonal matrix, whose diagonal elements are the eigenvalues of $\mathbf{\Psi}_l$. Using the least squares (LS), we can obtain the following formulation:

$$\min\|\Delta\mathbf{U}\|^2 \text{ s.t.} \mathbf{U}_{S_l} + \Delta\mathbf{U} = \mathbf{U}_{S_0}\mathbf{\Psi}_l. \tag{16}$$

From (16), we can derive

$$\begin{aligned} \mathbf{\Psi}_l &= \left(\mathbf{U}_{S_0}^H\mathbf{U}_{S_0}\right)^{-1}\mathbf{U}_{S_0}^H\mathbf{U}_{S_l} \\ &= \mathbf{U}_{S_0}^+\mathbf{U}_{S_l} \end{aligned}, \tag{17}$$

where $(\cdot)^+$ represents pseudo-inverse. Obviously, $\hat{\mathbf{\Phi}}_l$ can be estimated by performing the eigen-decomposition of $\mathbf{\Psi}_l$. Iterate through different array elements and repeat the aforementioned steps (17), the spatial manifold matrix $\hat{\mathbf{B}}$ can be estimated as follows:

$$\hat{\mathbf{B}} = \begin{bmatrix} \text{diag}^{-1}(\hat{\mathbf{\Phi}}_{-L}) \\ \vdots \\ \text{diag}^{-1}(\hat{\mathbf{\Phi}}_0) \\ \vdots \\ \text{diag}^{-1}(\hat{\mathbf{\Phi}}_L) \end{bmatrix} = \left[\hat{\mathbf{b}}(\theta_1, r_1), \ldots, \hat{\mathbf{b}}(\theta_K, r_K)\right], \tag{18}$$

where $\text{diag}^{-1}(\cdot)$ represents the row vector operation, and the estimation of spatial steering vector of k-th signal is $\hat{\mathbf{b}}(\theta_k, r_k) = \left[\hat{b}_{-L,k}, \ldots, \hat{b}_{0,k}, \ldots, \hat{b}_{L,k}\right]^T$.

3.2. Polarization Parameters Estimation

From (7), obviously, the FOC matrix $\mathbf{C}_l$ can be divided into

$$\mathbf{C}_l = \begin{bmatrix} \mathbf{B\Phi}_l\mathbf{P}_1\mathbf{C}_{4s}\mathbf{B}^H \\ \mathbf{B\Phi}_l\mathbf{P}_2\mathbf{C}_{4s}\mathbf{B}^H \end{bmatrix}, \tag{19}$$

where the first half of $\mathbf{C}_l$ corresponds to the horizontal component, while the second half corresponds to the vertical component. Therefore, $\mathbf{U}_{\mathrm{S}_l}$ can be further divided into

$$\mathbf{U}_{\mathrm{S}_l} = \left[\mathbf{U}_{\mathrm{S}_{l,1}}^T, \mathbf{U}_{\mathrm{S}_{l,2}}^T\right]^T, \tag{20}$$

where $\mathbf{U}_{\mathrm{S}_{l,1}}$ is formed by the first $2L+1$ rows of $\mathbf{U}_{\mathrm{S}_l}$ and $\mathbf{U}_{\mathrm{S}_{l,2}}$ formed by the last $2L+1$ rows. Sub-matrices $\mathbf{U}_{\mathrm{S}_{-L,1}}, \ldots, \mathbf{U}_{\mathrm{S}_{L,1}}$ and $\mathbf{U}_{\mathrm{S}_{-L,2}}, \ldots, \mathbf{U}_{\mathrm{S}_{L,2}}$ can be extracted from the signal subspace $\mathbf{U}_{\mathrm{S}}$, the horizontal and vertical component subspaces, $\mathbf{E}_h$ and $\mathbf{E}_v$, can then be constructed as

$$\mathbf{E}_h = \begin{bmatrix} \mathbf{U}_{\mathrm{S}_{-L,1}} \\ \vdots \\ \mathbf{U}_{\mathrm{S}_{0,1}} \\ \vdots \\ \mathbf{U}_{\mathrm{S}_{L,1}} \end{bmatrix}, \mathbf{E}_v = \begin{bmatrix} \mathbf{U}_{\mathrm{S}_{-L,2}} \\ \vdots \\ \mathbf{U}_{\mathrm{S}_{0,2}} \\ \vdots \\ \mathbf{U}_{\mathrm{S}_{L,2}} \end{bmatrix}. \tag{21}$$

By applying ESPRIT, we can rewrite (21) as

$$\mathbf{E}_h = \begin{bmatrix} \mathbf{B}\mathbf{\Phi}_{-L}\mathbf{P}_1 \\ \vdots \\ \mathbf{B}\mathbf{\Phi}_{L}\mathbf{P}_1 \end{bmatrix}\mathbf{T}, \mathbf{E}_v = \begin{bmatrix} \mathbf{B}\mathbf{\Phi}_{-L}\mathbf{P}_2 \\ \vdots \\ \mathbf{B}\mathbf{\Phi}_{L}\mathbf{P}_2 \end{bmatrix}\mathbf{T}. \tag{22}$$

With component matrices $\mathbf{E}_h$ and $\mathbf{E}_v$, we can obtain the following matrix:

$$\begin{aligned} \mathbf{D} &= \mathbf{E}_h^+ \mathbf{E}_v \\ &= \mathbf{T}^{-1}\mathbf{P}_1^{-1}\mathbf{P}_2\mathbf{T} = \mathbf{T}^{-1}\mathbf{P}\mathbf{T} \end{aligned}. \tag{23}$$

Substituting matrices $\mathbf{P}_1$ and $\mathbf{P}_2$ into (23), the polarized diagonal matrix $\mathbf{P}$ can be expressed as

$$\mathbf{P} = \mathrm{diag}\left[\cot\gamma_1 e^{-j\eta_1}, \ldots, \cot\gamma_K e^{-j\eta_K}\right]. \tag{24}$$

Like (15), $\hat{\mathbf{P}}$ can be estimated by performing eigen-decomposition of $\mathbf{D}$. As a result, the polarization parameters estimates of k-th source are given by

$$\begin{aligned} \hat{\gamma}_k &= \mathrm{arc}\cot\left|\hat{\mathbf{P}}_k\right| \\ \hat{\eta}_k &= -\mathrm{angle}\left(\hat{\mathbf{P}}_k\right) \end{aligned}, \tag{25}$$

where $\hat{\mathbf{P}}_k$ represents the k-th diagonal element of $\hat{\mathbf{P}}$.

3.3. Coarse Estimation

The angle$(\cdot)$ operator can extract the phase of $\hat{b}_{l,k}$ but introduces phase ambiguity when the spacing between two COLDs exceeds $\lambda/2$. To resolve this ambiguity, disambiguation operations utilizing the coarse reference estimates are employed to obtain the unambiguous phase $\hat{\tau}_{l,k}$ for fine estimation. Then, we impose structural constraints on the array configuration to obtain unambiguous reference values. Assuming there exist three sensors, whose positions are d_p, d_c, and d_q, respectively, the phase factor will be coarse but unambiguous, when the spacing of these three sensors satisfies $\left|d_p - d_c\right| \leqslant \lambda/2$ and $\left|d_q - d_c\right| \leqslant \lambda/2$. The relationship between these sensors can be established as

$$\begin{cases} \sqrt{r_k^2 + d_c^2 - 2r_k d_c \sin\alpha_k} - \sqrt{r_k^2 + d_p^2 - 2r_k d_p \sin\alpha_k} = \frac{\hat{\psi}_{c,p,k}}{2\pi/\lambda} \\ \sqrt{r_k^2 + d_c^2 - 2r_k d_c \sin\alpha_k} - \sqrt{r_k^2 + d_q^2 - 2r_k d_q \sin\alpha_k} = \frac{\hat{\psi}_{c,q,k}}{2\pi/\lambda} \end{cases}, \tag{26}$$

where $\hat{\psi}_{c,p,k} = \text{angle}\left(\hat{\mathbf{B}}(p,k)/\hat{\mathbf{B}}(c,k)\right)$ and $\hat{\psi}_{c,p,k} = \text{angle}\left(\hat{\mathbf{B}}(q,k)/\hat{\mathbf{B}}(c,k)\right)$. In this work, d_c represents the coordinates of the reference element, while $d_p = d_{-1}$ and $d_q = d_1$ denote the coordinates of its adjacent elements located at $\lambda/2$ on either side. Hence, from (18), the unambiguous phases of the k-th steering vector can be denoted as $\hat{\tau}_{-1,k}$ and $\hat{\tau}_{1,k}$. Then, the coarse estimates of θ_k and r_k can be given by

$$\begin{cases} \hat{r}_k^{(c)} & = \dfrac{d_1\left(\frac{\hat{\tau}_{-1,k}}{2\pi/\lambda}\right)^2 - d_{-1}\left(\frac{\hat{\tau}_{1,k}}{2\pi/\lambda}\right)^2 + d_{-1}d_1^2 - d_1 d_{-1}^2}{2d_1\frac{\hat{\tau}_{-1,k}}{2\pi/\lambda} - 2d_{-1}\frac{\hat{\tau}_{1,k}}{2\pi/\lambda}} \\ \hat{\theta}_k^{(c)} & = \arcsin\left(\dfrac{2\hat{r}_k^{(c)}\frac{\hat{\tau}_{-1,k}}{2\pi/\lambda} - \left(\frac{\hat{\tau}_{-1,k}}{2\pi/\lambda}\right)^2 + d_{-1}^2}{2\hat{r}_k^{(c)}d_{-1}}\right) \end{cases}. \tag{27}$$

3.4. Fine Estimation

Let $\hat{\phi}_{l,k}$ denote the ambiguous phase. According to the (2), there exists an integer-multiplicity set S_l relationship between $\hat{\phi}_{l,k}$ and $\hat{\tau}_{l,k}$, expressed as $S_l = \left\{\hat{\phi}_{l,k} + 2\pi n \middle|\ n \in \mathcal{Z}\right\}$. With (27), by using $\hat{\theta}_k^{(c)}$ and $\hat{r}_k^{(c)}$, we can construct the unambiguous phase reference values as follows:

$$\hat{\tau}_{l,k}^{co} = \frac{2\pi}{\lambda}\left(\hat{r}_k^{(c)} - \hat{r}_{l,k}^{(c)}\right). \tag{28}$$

Thus, the exact estimation $\hat{\tau}_{l,k}$ is the value in S_l that minimizes the deviation from the coarse estimate $\hat{\tau}_{l,k}^{co}$, i.e.,

$$\hat{\tau}_{l,k} = \arg\min_n \left|\tau_{l,k}^{co} - \hat{\phi}_{l,k} + 2\pi n\right|. \tag{29}$$

Furthermore, the phase term satisfies the following relationship with respect to the angle and range:

$$\frac{\tau_{l,k}}{2\pi/\lambda} = r_k - r_{l,k} = r_k - \sqrt{r_k^2 + d_l^2 - 2r_k d_l \sin\theta_k}. \tag{30}$$

Substituting the $\hat{\tau}_{-L,k}, \cdots, \hat{\tau}_{l,k}, \cdots, \hat{\tau}_{L,k}$ into (30) and simplifying, we can construct coefficient matrices which consist of $2L$ equation relations

$$\mathbf{V}_k = \begin{bmatrix} 2\frac{\hat{\tau}_{-L,k}}{2\pi/\lambda} & -2d_{-L} \\ \vdots & \vdots \\ 2\frac{\hat{\tau}_{L,k}}{2\pi/\lambda} & -2d_L \end{bmatrix}, \mathbf{W}_k = \begin{bmatrix} \left(\frac{\hat{\tau}_{-L,k}}{2\pi/\lambda}\right)^2 - d_{-L}^2 \\ \vdots \\ \left(\frac{\hat{\tau}_{L,k}}{2\pi/\lambda}\right)^2 - d_L^2 \end{bmatrix}, \tag{31}$$

where the unknown vector to be solved is denoted as $\mathbf{\Delta}_{f,k}^T = [r_k\ r_k \sin\theta_k]^T$. These vectors can be further rewritten as a matrix equation $\mathbf{V}_k \times \mathbf{\Delta}_{f,k} = \mathbf{W}_k$. By applying LS, the $\hat{\mathbf{\Delta}}_{f,k}$ can be solved as

$$\begin{aligned} \hat{\mathbf{\Delta}}_{f,k} &= \left(\mathbf{V}_k^H \mathbf{V}_k\right)^{-1} \mathbf{V}_k^H \mathbf{W}_k \\ &= \mathbf{V}_k^+ \mathbf{W}_k \end{aligned}. \tag{32}$$

Finally, the fine estimates of angle and range can be derived as

$$\begin{cases} \hat{r}_k^{fi} = \hat{\mathbf{\Delta}}_{f,k}(1) \\ \hat{\theta}_k^{fi} = \arcsin\left(\frac{\hat{\mathbf{\Delta}}_{f,k}(2)}{\hat{\mathbf{\Delta}}_{f,k}(1)}\right) \end{cases}. \tag{33}$$

3.5. Asymptotic Variance for Joint Parameters

The theoretical phase at the l-th sensor from the k-th source is given by

$$\vartheta_{l,k} = e^{j\frac{2\pi}{\lambda}\left(r_k - \sqrt{r_k^2 + d_l^2 - 2r_k d_l \sin\theta_k}\right)}, \tag{34}$$

while the corresponding polarization factor for the k-th source is

$$p_k = \cot\gamma_k e^{-j\eta_k}. \tag{35}$$

Additionally, the estimate $\hat{\vartheta}_{l,k}$ of $\vartheta_{l,k}$, is the (l,k)-th element of $\hat{\mathbf{B}}$, while $\hat{p}_k$ is obtained from the k-th diagonal element of $\hat{\mathbf{P}}$. Thus, the deviations between the true and estimated values are $\delta\vartheta_{l,k} = \hat{\vartheta}_{l,k} - \vartheta_{l,k}$ and $\delta p_k = \hat{p}_k - p_k$ [13].

For simplicity, the fine estimates of angle and range are defined as $\hat{\theta}_k$ and $\hat{r}_k$, whose deviations are defined as $\delta\theta_k = \hat{\theta}_k - \theta_k$ and $\delta r_k = \hat{r}_k - r_k$. In addition, the deviations of polarization angle and polarization phase difference are defined as $\delta\gamma_k = \hat{\gamma}_k - \gamma_k$ and $\delta\eta_k = \hat{\eta}_k - \eta_k$, respectively.

Using the first-order Taylor expansion for $\vartheta_{l,k}$ and p_k, we can obtain

$$\vartheta_{l,k} \approx \hat{\vartheta}_{l,k} + (\theta_k - \hat{\theta}_k)\frac{\partial\vartheta_{l,k}}{\partial\theta_k} + (r_k - \hat{r}_k)\frac{\partial\vartheta_{l,k}}{\partial r_k}, \tag{36}$$

$$p_k \approx \hat{p}_k + (\gamma_k - \hat{\gamma}_k)\frac{\partial p_k}{\partial\gamma_k} + (\eta_k - \hat{\eta}_k)\frac{\partial p_k}{\partial\eta_k}, \tag{37}$$

where the partial derivatives of $\vartheta_{l,k}$ to $\hat{\theta}_k$ and $\hat{r}_k$ are

$$\begin{aligned}
\frac{\partial\vartheta_{l,k}}{\partial\theta_k} &= \vartheta_{l,k} j\frac{2\pi}{\lambda}\left(\frac{r_k d_l \cos\theta_k}{\sqrt{r_k^2 + d_l^2 - 2r_k d_l \sin\theta_k}}\right), \\
\frac{\partial\vartheta_{l,k}}{\partial r_k} &= \vartheta_{l,k} j\frac{2\pi}{\lambda}\left(1 - \frac{2r_k - 2d_l \sin\theta_k}{2\sqrt{r_k^2 + d_l^2 - 2r_k d_l \sin\theta_k}}\right),
\end{aligned} \tag{38}$$

and

$$\begin{aligned}
\frac{\partial p_k}{\partial\gamma_k} &= -\frac{1}{\sin^2\gamma_k}e^{-j\eta_k}, \\
\frac{\partial p_k}{\partial\eta_k} &= -j\cot\gamma_k e^{-j\eta_k}.
\end{aligned} \tag{39}$$

Substituting (38) and (39) into (36) and (37), respectively, the final expansion yields

$$j\frac{2\pi r_k d_l}{\lambda r_{l,k}}\cos\theta_k \vartheta_{l,k}\delta\theta_k + j\frac{2\pi(r_{l,k} - r_k + d_l \sin\theta_k)}{\lambda r_{l,k}}\vartheta_{l,k}\delta r_k = \delta\vartheta_{l,k}, \tag{40}$$

$$-\frac{1}{\sin^2\gamma_k}e^{-j\eta_k}\delta\gamma_k - j\cot\gamma_k e^{-j\eta_k}\delta\eta_k = \delta p_k. \tag{41}$$

With the same principle as (31), $2L$ equations can be listed, which can than be rewritten in matrix form

$$\mathbf{Z}_k = \begin{bmatrix} jv_{-L,k}\vartheta_{-L,k} & j\omega_{-L,k}\vartheta_{-L,k} \\ \vdots & \vdots \\ jv_{L,k}\vartheta_{L,k} & j\omega_{L,k}\vartheta_{L,k} \end{bmatrix}, \mathbf{\Lambda}_k = \begin{bmatrix} \delta\vartheta_{-L,k} \\ \vdots \\ \delta\vartheta_{L,k} \end{bmatrix}, \tag{42}$$

where

$$v_{l,k} = \frac{2\pi r_k d_l}{\lambda r_{l,k}} \cos\theta_k, \omega_{l,k} = \frac{2\pi(r_{l,k} - r_k + d_l \sin\theta_k)}{\lambda r_{l,k}}, \tag{43}$$

and the unknown vector to be solved is $\mathbf{\Gamma}_k = [\delta\theta_k\ \delta r_k]^T$, which satisfies $\mathbf{Z}_k \times \mathbf{\Gamma}_k = \mathbf{\Lambda}_k$. By applying LS, we have the following results:

$$\mathbf{\Gamma}_k = \left(\mathbf{Z}_k^H \mathbf{Z}_k\right)^{-1} \mathbf{Z}_k^H \mathbf{\Lambda}_k = \mathbf{Z}_k^+ \mathbf{\Lambda}_k, \tag{44}$$

Thus, the closed-form variance expressions can be given by

$$\begin{aligned} E\left\{(\delta\theta_k)^2\right\} &= E\left\{\left[\mathbf{Z}_k^+ \mathbf{\Lambda}_k (\mathbf{Z}_k^+ \mathbf{\Lambda}_k)^T\right]_1\right\}, \\ E\left\{(\delta r_k)^2\right\} &= E\left\{\left[\mathbf{Z}_k^+ \mathbf{\Lambda}_k (\mathbf{Z}_k^+ \mathbf{\Lambda}_k)^T\right]_2\right\}. \end{aligned} \tag{45}$$

where $[\cdot]_1$ represents the first diagonal element, and $[\cdot]_2$ represents the second diagonal element.

According to (41), $\delta\gamma_k$ and $\delta\eta_k$ correspond to the real and imaginary parts of δp_k, respectively. By exploiting the conjugate symmetry, the closed-form expressions for $\delta\gamma_k$ and $\delta\eta_k$ are

$$\begin{aligned} \delta\gamma_k &= \frac{(\delta p_k e^{j\eta_k} + \delta p_k^* e^{-j\eta_k}) \sin^2\gamma_k}{-2}, \\ \delta\eta_k &= \frac{\delta p_k e^{j\eta_k} - \delta p_k^* e^{-j\eta_k}}{-2j\cot\gamma_k}, \end{aligned} \tag{46}$$

and the variances of $\delta\gamma_k$ and $\delta\eta_k$ can be given by

$$\begin{cases} E\left\{(\delta\gamma_k)^2\right\} = \frac{E\left\{\left(\delta p_k e^{j\eta_k} + \delta p_k^* e^{-j\eta_k}\right)^2\right\} \sin^4\gamma_k}{4} \\ E\left\{(\delta\eta_k)^2\right\} = \frac{E\left\{\left(\delta p_k e^{j\eta_k} - \delta p_k^* e^{-j\eta_k}\right)^2\right\}}{-4\cot^2\gamma_k} \end{cases}. \tag{47}$$

The steps of the proposed method are summarized in Table 1.

Table 1. Algorithmic steps.

Step	Operation
1	Construct $\mathbf{C}_{-L}, \cdots, \mathbf{C}_0, \cdots, \mathbf{C}_L$ via (9) and the long matrix Ξ can be obtain via (10).
2	Compute the covariance matrix $\mathbf{R}$ and perform eigen-decomposition to obtain signal subspace $\mathbf{U}_S$.
3	Using the LS and eigen-decomposition, we can obtain $\hat{\mathbf{\Phi}}_{-L}, \cdots, \hat{\mathbf{\Phi}}_0, \cdots, \hat{\mathbf{\Phi}}_L$ to construct $\hat{\mathbf{B}}$.
4	Extract $\mathbf{U}_{S_{-L,1}}, \ldots, \mathbf{U}_{S_{L,1}}$ and $\mathbf{U}_{S_{-L,2}}, \ldots, \mathbf{U}_{S_{L,2}}$ from $\mathbf{U}_S$ to construct $\mathbf{E}_h$ and $\mathbf{E}_v$. Matrix $\mathbf{P}$ is obtained by performing eigen-decomposition on matrix $\mathbf{D}$, where matrix $\mathbf{D}$ can be obtain from (33). From the diagonal element of $\mathbf{P}$, we have the estimates $\hat{\gamma}_1, \cdots, \hat{\gamma}_k, \cdots, \hat{\gamma}_K$ and $\hat{\eta}_1, \cdots, \hat{\eta}_k, \cdots, \hat{\eta}_K$.

5	**for** $k \in [1, K]$ **do**
	(1) The closed-form coarse estimation $\hat{r}_k^{(c)}$ and $\hat{\theta}_k^{(c)}$ can be obtained via (26) and (27).
	(2) Via (28) and (29), we can disambiguate the phase to construct matrices $\mathbf{V}_k$ and $\mathbf{W}_k$.
	(3) Based on LS method, the fine estimation $\hat{r}_k^{fi}$ and $\hat{\theta}_k^{fi}$ can be obtained by solving (33).
	End for

3.6. Computational Complexity

In this section, we analyze the computational complexity of the proposed method, focusing on (a) $2L+1$ FOC matrix construction requires $\mathcal{O}\left\{4(2L+1) \times (2L+1)^2 N\right\}$ flops;

(b) the complexity of computing the covariance matrix $\mathbf{R}$ requires $\mathcal{O}\{8(2L+1)^5\}$; (c) the eigen-decomposition of the covariance matrix requires $\mathcal{O}\{8(2L+1)^6\}$ multiplication operations; (d) the complexity of manifold matrix estimation requires $\mathcal{O}\{4(2L+1)K^2+2K^2\}$; (e) the complexity in computing $\mathbf{D}$ is $\mathcal{O}\{2(2L+1)K^2\}$. Overall, the proposed method is approximately $\mathcal{O}\{4(2L+1)^3N+8(2L+1)^5+8(2L+1)^6+(12L+8)K^2\}$. The computational complexity of the proposed algorithm and that of other comparative methods are summarized in Table 2, where n_θ, n_r, n_γ, n_η represent the number of searches for DOA, range, polarization angle, and polarization phase difference. As shown in Table 2, the proposed method showed better efficiency than FR-RARE and G-ESPRIT without peak search, though requiring more operations than He's method.

Table 2. Computational complexity of all compared algorithms.

Method	Complexity
Proposed	$\mathcal{O}\{4(2L+1)^3N+8(2L+1)^5+8(2L+1)^6+(12L+8)K^2\}$
FR-RARE [22]	$\mathcal{O}\{4(2L+1)^2N+8(2L+1)^3+4(L+1)(4L+2-K)(2L+1)Kn_\theta$ $+(2L+1)(32L+16-4K+8)Kn_r+2(2L+1)(4L+2-K)n_\gamma n_\eta\}$
G-ESPRIT [23]	$\mathcal{O}\{4(2L+1)^2N+8(2L+1)^3+32LK^2n_\theta+4(2L+1)(4L+2-K)n_r\}$
He [24]	$\mathcal{O}\{4(2L+1)^2N+8(2L+1)^3+4(2L+1)^2K+2(2L+1)K^2\}$

3.7. CRB Derivation

This section derives the closed-form deterministic Cramér–Rao Bound (CRB) expressions for jointly estimating the DOA, range, and polarization parameters under exact near-field polarization conditions. We define the real-valued vector $\boldsymbol{\varepsilon}=\left[\boldsymbol{\theta}^T\ \boldsymbol{r}^T\ \boldsymbol{\eta}^T\ \boldsymbol{\gamma}^T\right]^T$, where $\boldsymbol{\theta}=[\theta_1,\theta_2,\cdots,\theta_K]^T$, $\boldsymbol{r}=[r_1,r_2,\cdots,r_K]^T$, $\boldsymbol{\eta}=[\eta_1,\eta_2,\cdots,\eta_K]^T$ and $\boldsymbol{\gamma}=[\gamma_1,\gamma_2,\cdots,\gamma_K]^T$. Thus, the CRB for the parameters in $\boldsymbol{\varepsilon}$ is given by [24]

$$\mathbf{CRB}(\boldsymbol{\varepsilon})=\frac{\sigma_n^2}{2N}\left\{\Re\left[\left(\mathbf{Y}^H\mathbf{P}_{\boldsymbol{\Theta}}^{\perp}\mathbf{Y}\right)\oplus\left(\mathbf{1}_4\odot\mathbf{1}_4^T\otimes\mathbf{R}_s^T\right)\right]\right\}^{-1}. \quad (48)$$

where $\mathbf{Y}=[\mathbf{Y}_\theta,\mathbf{Y}_r,\mathbf{Y}_\gamma,\mathbf{Y}_\eta]$, $\mathbf{Y}_\theta=\left[\frac{\partial\boldsymbol{\Theta}}{\partial\theta_1},\cdots,\frac{\partial\boldsymbol{\Theta}}{\partial\theta_K}\right]$, $\mathbf{Y}_r=\left[\frac{\partial\boldsymbol{\Theta}}{\partial r_1},\cdots,\frac{\partial\boldsymbol{\Theta}}{\partial r_K}\right]$, $\mathbf{Y}_\gamma=\left[\frac{\partial\boldsymbol{\Theta}}{\partial\gamma_1},\cdots,\frac{\partial\boldsymbol{\Theta}}{\partial\gamma_K}\right]$, $\mathbf{Y}_\eta=\left[\frac{\partial\boldsymbol{\Theta}}{\partial\eta_1},\cdots,\frac{\partial\boldsymbol{\Theta}}{\partial\eta_K}\right]$, $\mathbf{P}_{\boldsymbol{\Theta}}^{\perp}=\mathbf{I}_{2(2L+1)}-\boldsymbol{\Theta}\left(\boldsymbol{\Theta}^H\boldsymbol{\Theta}\right)^{-1}\boldsymbol{\Theta}^H$.

4. Simulation Results

Simulation experiments were performed to evaluate the performance of the proposed method compared with the existing G-ESPRIT method [22], FR-RARE method [23], He's method [24], and the CRB [24] for an exact NF polarization scenario.

4.1. Experiment 1: RMSE vs. SNR

In this section, a symmetric linear COLD array consisting of 9 antenna elements was considered, with uniform spacing $\lambda/4$. Two uncorrelated NF sources arrived at the COLD array, with parameters $(-15^\circ, 5.5\lambda, 20^\circ, -20^\circ)$ and $(20^\circ, 6\lambda, 30^\circ, 40^\circ)$ and the number of snapshots was set to $N=2000$. The root-mean-square error (RMSE) was utilized to evaluate the performance, defined as

$$\text{RMSE}=\sqrt{\frac{1}{500K}\sum_{k=1}^{K}\sum_{i=1}^{500}(x_k-\hat{x}_{i,k})^2}, \quad (49)$$

where x_k stands for the true value of parameters θ_k, r_k, γ_k, η_k and $\hat{x}_{i,k}$ represents the estimate of the $\hat{\theta}_{i,k}$, $\hat{r}_{i,k}$, $\hat{\gamma}_{i,k}$, $\hat{\eta}_{i,k}$ at the i-th run in 500 Monte-Carlo trials.

Figure 2 shows the RMSE curves of the spatial and polarization parameters, when the SNR varied from 0 dB to 40 dB. It can be seen that the estimation performance of both the proposed algorithm and He's method improved as the SNR increased. As an amplitude-attenuation based method, He's method exhibited limited estimation accuracy, suffering from mismatch under the low SNR. Among all the algorithms, the proposed algorithm more closely approximated the CRB and could exploit the whole aperture. In addition, due to the model approximation, G-ESPRIT and FR-RARE showed performance "saturation".

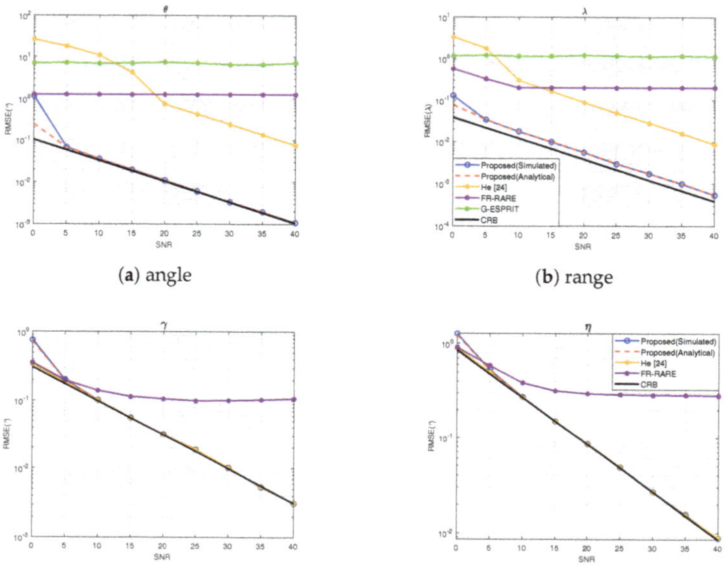

(**a**) angle (**b**) range

(**c**) polarization angle (**d**) polarization phase difference

Figure 2. RMSE vs. SNR.

4.2. Experiment 2: Scatter for Spatial and Polarization

In this section, we derived the maximum number of identifiable NF sources. Based on an analysis of the rank of the manifold matrix, the theoretical limit was established when the matrix satisfied the full-rank condition. Consider a 9-element COLD array with antenna positions $[-2, -1.6, -1.1, -0.5, 0, 0.5, 1.3, 1.5, 2.2]$. Assume that 8 uncorrelated NF sources arrive at the array with locations $(-15°, 5.5\lambda, 20°, -20°)$, $(-5°, 6\lambda, 30°, 40°)$, $(5°, 6.5\lambda, 45°, 30°)$, $(20°, 7.5\lambda, 50°, 25°)$, $(25°, 5.8\lambda, 25°, -10°)$, $(30°, 6\lambda, 20°, 42°)$, $(35°, 6.3\lambda, 40°, 10°)$, and $(40°, 6.8\lambda, 30°, 50°)$. With the SNR fixed at 25 dB and the number of snapshots be 2000, Figure 3 presents scatter plots of joint parameters in a 2-D plane based on 500 Monte Carlo runs. As shown in Figure 3, the 4-D parameters were all closely distributed near the true values, which were automatically paired.

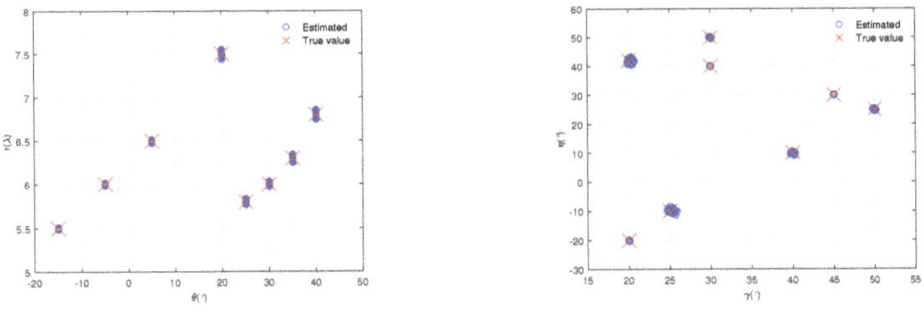

(**a**) spatial parameters. (**b**) polarization parameters

Figure 3. Scatter plots of 4-D parameters estimation.

4.3. *Experiment 3: RMSE vs. Path Loss Exponent*

In this section, we tested the algorithm performance with difference path loss exponents. The location of the sources and other parameters were set the same as in Experiment 1, and the path loss exponent varied from 2 to 6. As shown in Figure 4, the RMSE was approximately constant with the exponential increase in path loss. The reasons for the performance saturation of He's method, G-ESPRIT, and FR-RARE were the same as in Experiment 1.

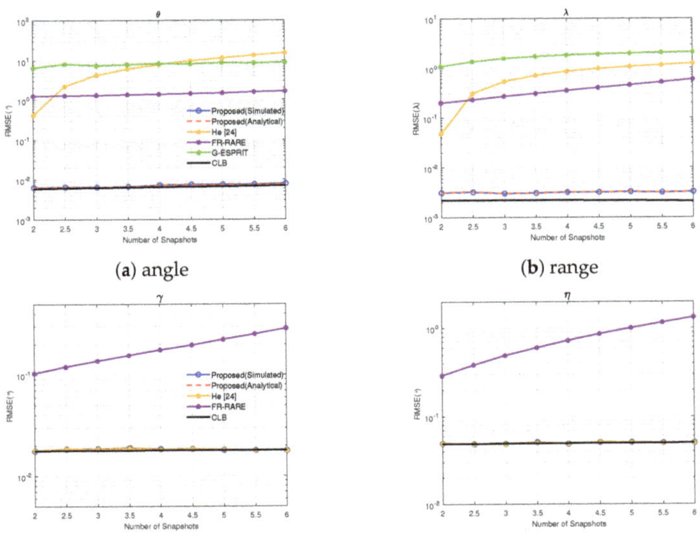

(**a**) angle (**b**) range

(**c**) polarization angle (**d**) polarization phase difference

Figure 4. RMSE vs. path loss exponent.

4.4. *Experiment 4: Detection Capability for Closely Spaced Targets*

In this section, we considered a scenario where two closely spaced impinging signals had the same DOA or same range. In Figure 5a, two sources had the same DOA parameterized by $(20^\circ, 5.5\lambda, 20^\circ, -20^\circ)$, and $(20^\circ, 5.5\lambda + \delta\lambda, 30^\circ, 40^\circ)$, in Figure 5b, the two sources had the same range parameterized by $(20^\circ, 5.5\lambda, 20^\circ, -20^\circ)$, and $(20^\circ + \delta\theta, 5.5\lambda, 30^\circ, 40^\circ)$, and the other parameters were set as same as Experiment 2. Successful detection was defined as a polarization angle RSME between the estimated $\hat{\gamma}$ and true value γ less than 1°. The probability of successful detection (PSD) was defined as a successful detection trial under 500 Monte-Carlo runs. As shown in Figure 5, when the DOA interval was larger than 0.5° or the range interval was larger than 0.4λ, the proposed method offered 100%

PSD for polarization angle; that is, we could refer to polarization information to distinguish two closely spaced impinging signals.

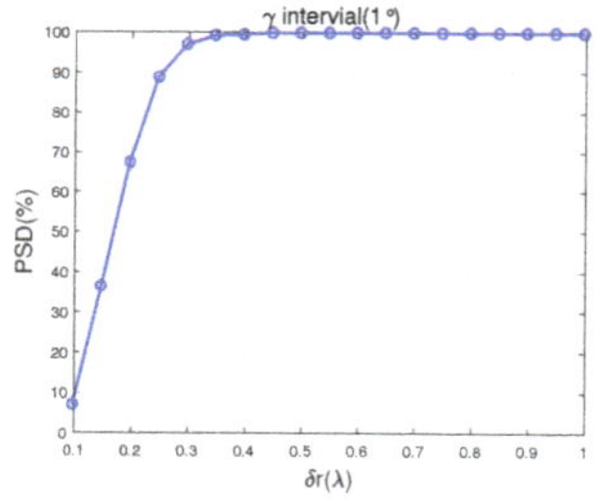

(**a**) two sources have the same DOA

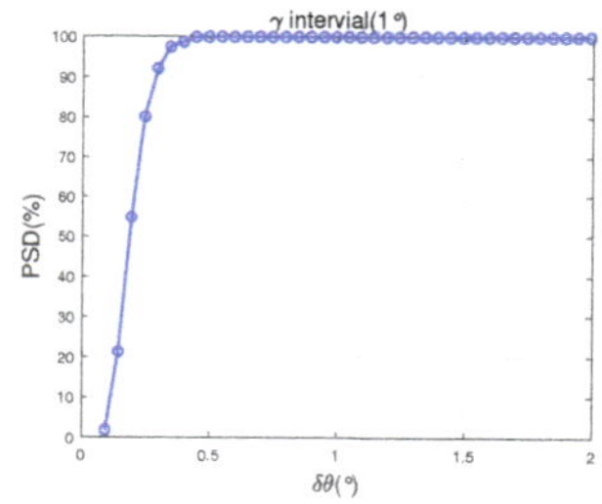

(**b**) two sources have the same range

Figure 5. PSD vs. parameter interval.

4.5. Experiment 5: RMSE vs. Correlation Coefficient

Assume that there are two correlated sources parameterized by $(10°, 5.5\lambda, 15°, 45°)$ and $(22°, 5.8\lambda, 35°, 90°)$, the other parameters were set the same as in Experiment 2, where correlation coefficient varied from 0.1 to 0.9. As can be seen from Figure 6, the RMSE of all spatial and polarization parameters increased with the increase in the correlation coefficient.

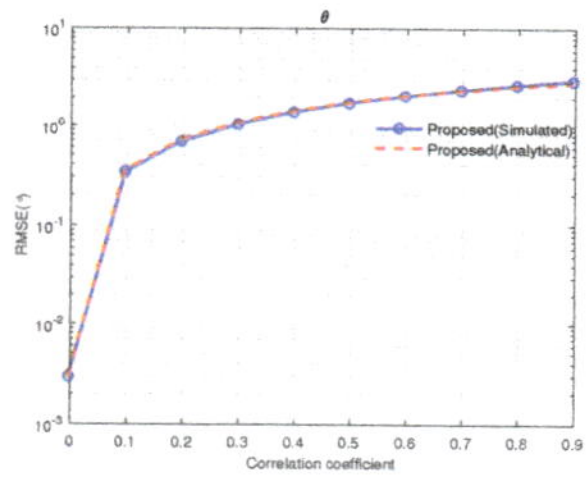

(**a**) angle

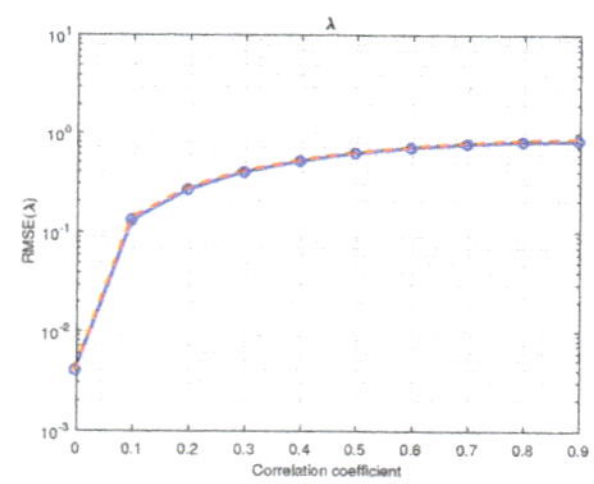

(**b**) range

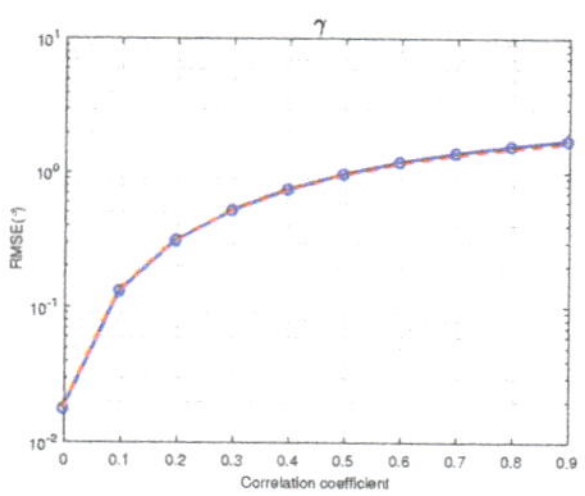

(**c**) polarization angle

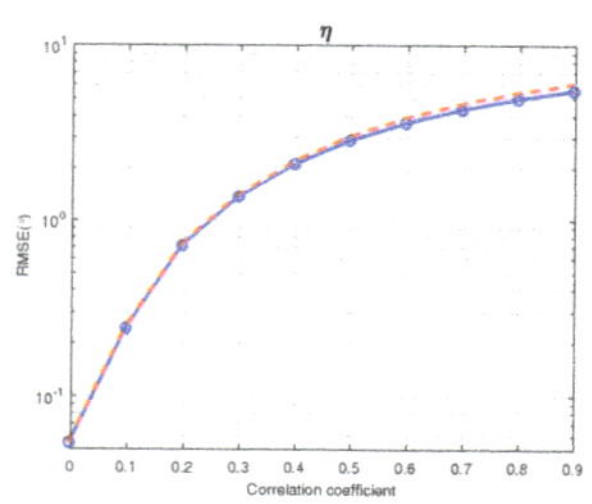

(**d**) polarization phase difference

Figure 6. RMSE vs. correlation coefficient.

4.6. Experiment 6: RMSE vs. Snapshots

In this experiment, we evaluated the performance by varying the number of snapshots from 400 to 5000, the SNR was fixed as 25 dB, with the same source parameters as in Experiment 1. The RMSE of the 4-D parameters vs. the number of snapshots is shown in Figure 7. The results demonstrate that the RMSE of both the proposed algorithm and He's method decreased with increasing snapshots, while our method had a better performance. In contrast, G-ESPRIT and FR-RARE were always in a performance saturation state.

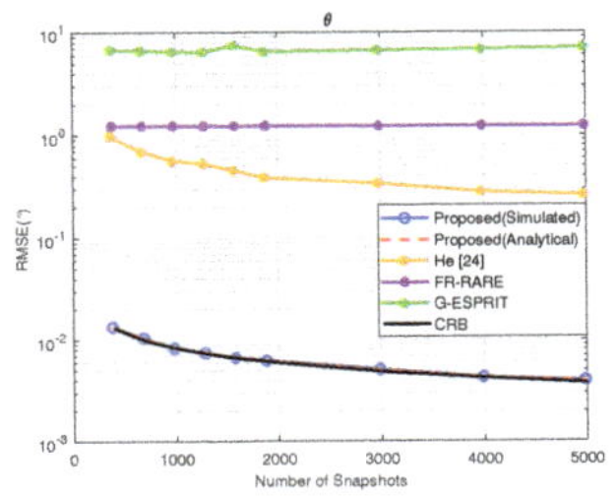

(**a**) angle

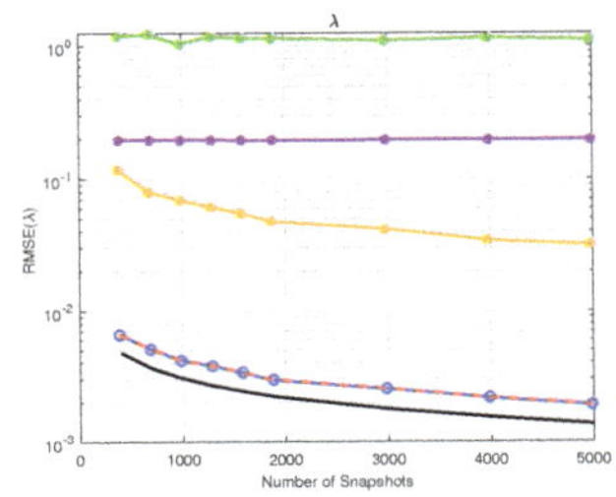

(**b**) range

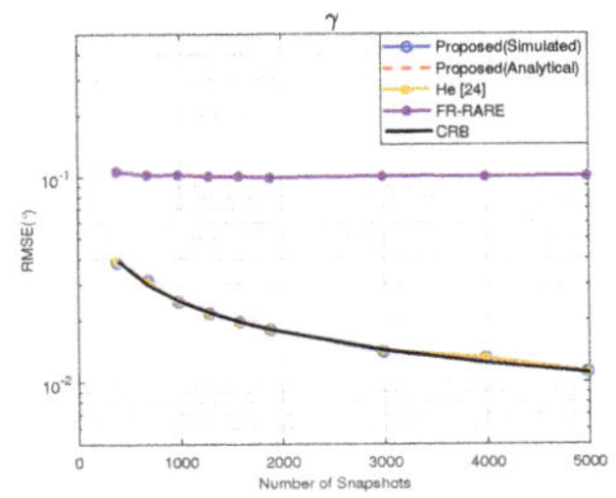

(**c**) polarization angle

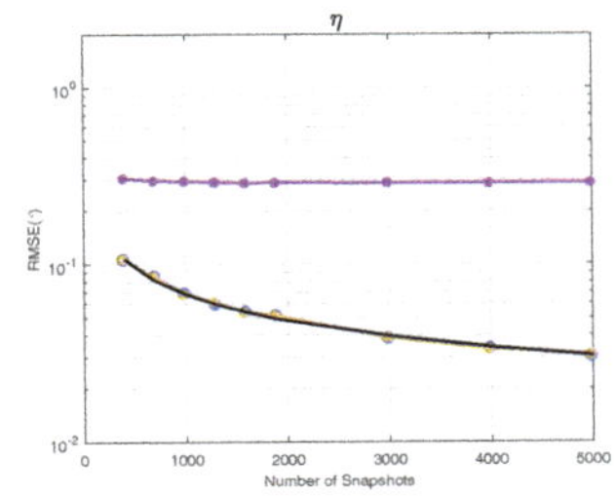

(**d**) polarization phase difference

Figure 7. RMSE vs. the number of snapshots.

5. Conclusions

The primary contribution of this paper is source localization under an exact NF polarization scenario. With a COLD array, a multiple-cumulant-matrices-based method was proposed for joint 4-D parameter estimation, formulated via least squares principles, with neither approximation nor aperture loss. After that, the asymptotic variances of 4-D parameters were derived. Simulation results demonstrated that the algorithm had superior performance under the exact NF polarization scenario.

Author Contributions: Conceptualization, J.Z.; methodology, J.Z., H.M. and Z.L.; software, J.Z. and Z.L.; validation, J.Z. and Z.L.; formal analysis, J.Z. and H.W.; investigation, J.Z. and Z.L.; resources, W.L.; data curation, J.Z. and H.M.; writing—original draft preparation, J.Z. and H.C.; writing—review and editing, H.C. and W.L.; visualization, H.C.; supervision, H.C.; project administration, H.C.; funding acquisition, H.C. and W.L. All authors have read and agreed to the published version of the manuscript.

Funding: This research was supported by Zhejiang Provincial Natural Science Foundation of China under Grant No. LY23F010003, and by the "Pioneer" and "Leading Goose" R & D Program of Zhejiang Province under Grant 2024C01105, and by the National Natural Science Foundation of China under Grant 62001256.

Institutional Review Board Statement: Not applicable.

Informed Consent Statement: Not applicable.

Data Availability Statement: The data used to support the findings of this study are available from the corresponding author upon request.

Acknowledgments: We sincerely thank all the reviewers for their patient reviews of and valuable suggestions for this article.

Conflicts of Interest: We declare that we have no financial and personal relationships with other people or organizations that can inappropriately influence our work, there is no professional or other personal interests of any nature or kind in any product, service, and/or company that could be construed as influencing the position presented in, or the review of, the manuscript entitled.

References

1. Sakhnini, A.; De Bast, S.; Guenach, M.; Bourdoux, A.; Sahli, H.; Pollin, S. Near-Field Coherent Radar Sensing Using a Massive MIMO Communication Testbed. *IEEE Trans. Wirel. Commun.* **2022**, *21*, 6256–6270. [CrossRef]
2. Ali, E.; Balfaqih, M.; Shglouf, I.; Alrayes, M.; Balfagih, Z. Two-Dimensional DOA Zero-Forcing Precoding for Enhanced Spectral Efficiency in Cell-Free Massive MIMO. In Proceedings of the 2025 22nd International Learning and Technology Conference, Jeddah, Saudi Arabia, 15–16 January 2025; Volume 22, pp. 164–169. [CrossRef]
3. Chen, H.; Fang, J.; Wang, W.; Liu, W.; Tian, Y.; Wang, Q.; Wang, G. Near-Field Target Localization for EMVS-MIMO Radar with Arbitrary Configuration. *IEEE Trans. Aerosp. Electron. Syst.* **2024**, *60*, 5406–5417. [CrossRef]
4. Weißer, F.; Utschick, W. DoA-Aided Channel Estimation for One-Bit Quantized Systems. In Proceedings of the 2025 14th International ITG Conference on Systems, Communications and Coding (SCC), Karlsruhe, Germany, 10–13 March 2025; pp. 1–6. [CrossRef]
5. Elbir, A.M.; Celik, A.; Eltawil, A.M. NEAT-MUSIC: Auto-Calibration of DOA Estimation for Terahertz-Band Massive MIMO Systems. *IEEE Wirel. Commun. Lett.* **2024**, *13*, 451–455. [CrossRef]
6. Buzzi, S.; Grossi, E.; Lops, M.; Venturino, L. Foundations of MIMO Radar Detection Aided by Reconfigurable Intelligent Surfaces. *IEEE Trans. Signal Process.* **2022**, *70*, 1749–1763. [CrossRef]
7. Chen, H.; Yi, Z.; Jiang, Z.; Liu, W.; Tian, Y.; Wang, Q.; Wang, G. Spatial-Temporal-Based Underdetermined Near-Field 3-D Localization Employing a Nonuniform Cross Array. *IEEE J. Sel. Top. Signal Process.* **2024**, *18*, 561–571. [CrossRef]
8. Chen, J.; Hudson, R.; Yao, K. Maximum-likelihood source localization and unknown sensor location estimation for wideband signals in the near-field. *IEEE Trans. Signal Process.* **2002**, *50*, 1843–1854. [CrossRef]
9. Gao, F.; Gershman, A. A generalized ESPRIT approach to direction-of-arrival estimation. *IEEE Signal Process. Lett.* **2005**, *12*, 254–257. [CrossRef]
10. Huang, Y.D.; Barkat, M. Near-field multiple source localization by passive sensor array. *IEEE Trans. Antennas Propag.* **1991**, *39*, 968–975. [CrossRef]
11. He, J.; Swamy, M.N.S.; Ahmad, M.O. Efficient Application of MUSIC Algorithm Under the Coexistence of Far-Field and Near-Field Sources. *IEEE Trans. Signal Process.* **2012**, *60*, 2066–2070. [CrossRef]
12. Zaman, F.; Akbar, S.; Akhtar, Z.; Phanomchoeng, G. Efficient Technique for Joint Parameter Estimation of Mixed Far Field Targets and Near Field Jammers in Phased Array Radar. *IEEE Access* **2025**, *13*, 55887–55898. [CrossRef]
13. Yuen, N.; Friedlander, B. Performance analysis of higher order ESPRIT for localization of near-field sources. *IEEE Trans. Signal Process.* **1998**, *46*, 709–719. [CrossRef]
14. Molaei, A.M.; del Hougne, P.; Fusco, V.; Yurduseven, O. Efficient Joint Estimation of DOA, Range and Reflectivity in Near-Field by Using Mixed-Order Statistics and a Symmetric MIMO Array. *IEEE Trans. Veh. Technol.* **2022**, *71*, 2824–2842. [CrossRef]
15. Liang, J.; Liu, D. Passive Localization of Mixed Near-Field and Far-Field Sources Using Two-stage MUSIC Algorithm. *IEEE Trans. Signal Process.* **2010**, *58*, 108–120. [CrossRef]
16. Friedlander, B. Localization of Signals in the Near-Field of an Antenna Array. *IEEE Trans. Signal Process.* **2019**, *67*, 3885–3893. [CrossRef]
17. Fang, J.; Chen, H.; Liu, W.; Yang, S.; Yuen, C.; So, H.C. Three-Dimensional Localization of Mixed Near-Field and Far-Field Sources Based on a Unified Exact Propagation Model. *IEEE Trans. Signal Process.* **2025**, *73*, 245–258. [CrossRef]
18. He, J.; Li, L.; Shu, T.; Truong, T.K. Mixed Near-Field and Far-Field Source Localization Based on Exact Spatial Propagation Geometry. *IEEE Trans. Veh. Technol.* **2021**, *70*, 3540–3551. [CrossRef]
19. Jiang, Z.; Chen, H.; Liu, W.; Wang, G. 3-D temporal-spatial-based near-field source localization considering amplitude attenuation. *Signal Process.* **2022**, *201*, 108735. [CrossRef]
20. Jin, L.; Chen, H.; Fang, J.; Liu, W.; Tian, Y.; Wang, G. Fourth-Order Cumulant Based 3-D Near-Field Underdetermined Parameter Estimation With Exact Spatial Propagation Model. In Proceedings of the ICASSP 2025—2025 IEEE International Conference on Acoustics, Speech and Signal Processing (ICASSP), Hyderabad, India, 6–11 April 2025; pp. 1–5. [CrossRef]
21. He, J.; Shu, T.; Li, L.; Truong, T.K. Mixed Near-Field and Far-Field Localization and Array Calibration With Partly Calibrated Arrays. *IEEE Trans. Signal Process.* **2022**, *70*, 2105–2118. [CrossRef]
22. He, J.; Ahmad, M.O.; Swamy, M.N.S. Near-Field Localization of Partially Polarized Sources with a Cross-Dipole Array. *IEEE Trans. Aerosp. Electron. Syst.* **2013**, *49*, 857–870. [CrossRef]
23. Tao, J.W.; Liu, L.; Lin, Z.Y. Joint DOA, Range, and Polarization Estimation in the Fresnel Region. *IEEE Trans. Aerosp. Electron. Syst.* **2011**, *47*, 2657–2672. [CrossRef]
24. He, J.; Li, L.; Shu, T. Near-Field Parameter Estimation for Polarized Source Using Spatial Amplitude Ratio. *IEEE Commun. Lett.* **2020**, *24*, 1961–1965. [CrossRef]

25. Mendel, J. Tutorial on higher-order statistics (spectra) in signal processing and system theory: Theoretical results and some applications. *Proc. IEEE* **1991**, *79*, 278–305. [CrossRef]
26. Ma, H.; Tao, H.; Xie, J.; Cheng, X. Parameters Estimation of 3-D Near-Field Sources with Arbitrarily Spaced Electromagnetic Vector-Sensors. *IEEE Commun. Lett.* **2022**, *26*, 2764–2768. [CrossRef]

sensors

MDPI

Communication

FDA-MIMO Radar Rapid Target Localization via Reconstructed Reduce Dimension Rooting

Cheng Wang [1,2], Zhi Zheng [3,*] and Wen-Qin Wang [2]

[1] Yangtze Delta Region Institute (Huzhou), University of Electronic Science and Technology of China, Huzhou 313002, China; chengw@csj.uestc.edu.cn

[2] School of Electronic Science and Engineering, University of Electronic Science and Technology of China, Chengdu 611731, China; wqwang@uestc.edu.cn

[3] Yangtze Delta Region Institute (Quzhou), University of Electronic Science and Technology of China, Quzhou 324000, China

* Correspondence: zz@uestc.edu.cn

Abstract: Frequency diversity array–multiple-input multiple-output (FDA-MIMO) radar realizes an angle- and range-dependent system model by adopting a slight frequency offset between adjacent transmitter sensors, thereby enabling potential target localization. This paper presents FDA-MIMO radar-based rapid target localization via the reduction dimension root reconstructed multiple signal classification (RDRR-MUSIC) algorithm. Firstly, we reconstruct the two-dimensional (2D)-MUSIC spatial spectrum function using the reconstructed steering vector, which involves no coupling of direction of arrival (DOA) and range. Subsequently, the 2D spectrum peaks search (SPS) is converted into one-dimensional (1D) SPS to reduce the computational complexity using a reduction dimension transformation. Finally, we conduct polynomial root finding to further eliminate computational costs, in which DOA and range can be rapidly estimated without performance degradation. The simulation results validate the effectiveness and superiority of the proposed RDRR-MUSIC algorithm over the conventional 2D-MUSIC algorithm and reduced-dimension (RD)-MUSIC algorithm.

Keywords: FDA-MIMO radar; rapid target localization; spectrum reconstruction; RDRR-MUSIC

1. Introduction

The multiple-input multiple-output (MIMO) radar fully utilizes space diversity to improve direction estimation accuracy, communication capacity, and interference suppression capability, which has been widely adopted in sonar, satellite navigation, and mobile communications [1,2]. However, the MIMO radar system model is independent of the target range, which limits the applications of the MIMO radar. Fortunately, the frequency diversity array (FDA) employs a slight frequency offset between adjacent transmitter sensors to realize the angle- and range-dependent beam pattern [3–5], which enjoys the potential of joint angle and range estimation. In order to fully utilize the advantages of the MIMO radar and FDA, a new radar system was investigated in [6–8], which merged the FDA and MIMO radar into a hybrid named the FDA-MIMO radar. The FDA-MIMO radar significantly improves the degrees of freedom (DOFs) of target localization, which has drawn much research attention. However, the ambiguous estimates caused by the coupling of the angle and range limit the applications of the FDA-MIMO radar. Researchers have proposed several schemes to solve this problem.

A double pulses scheme has been proposed to solve the coupling of the direction of arrival (DOA) and range [9], in which a pulse with zero frequency offset is utilized to estimate the DOA, and another pulse with non-zero frequency offset is exploited to estimate the range without coupling. However, the double pulses scheme adopts the maximum likelihood (ML) algorithm to estimate the DOA and range, which involves a high computational burden. In [10], a transmitter subaperture-based scheme was studied, in which the transmitter is divided into several subtransmitters to erase the coupling, and the DOA and range are successively estimated. However, the DOA and range are estimated by searching the 2D spatial domain, which restricts the practical uses of this transmitter subaperture-based scheme. Estimation of signal parameters via rotational invariance techniques (ESPRIT)-based algorithms [11–13] utilize the rotational invariance property of the structure to rapidly estimate the DOA and range without coupling, but the estimation accuracy needs to be further improved. The conventional 2D-MUSIC algorithm [14] can accurately estimate DOA and range using the 2D spectrum peaks search (SPS), which avoids an extra decoupling operation, but the computational complexity increases exponentially with an increasing number of search grids. The RD-MUSIC algorithm [15] exploits the reduction dimension transformation (RDT) to decouple DOA and range, and thereby reduces the computational complexity, but the DOA information is not fully utilized, resulting in DOA estimation performance degradation. Meanwhile, the error conduction caused by the extra decoupling operation also deteriorates the range estimation performance. In general, the algorithms mentioned above all require a trade-off between estimation reliability and estimation effectiveness.

In this paper, we propose the reduction dimension root reconstructed MUSIC (RDRR-MUSIC) algorithm to rapidly estimate DOA and range. We first reconstruct the 2D-MUSIC spatial spectrum function (SSF), utilizing the reconstructed steering vector without coupling the DOA and range. Then, we convert 2D SPS into 1D SPS using RDT; therefore, the computational complexity is significantly reduced. Finally, 1D SPS is transformed into polynomial root finding (PRF) to further reduce the computational complexity, and the DOA and range are rapidly estimated, but with no performance degradation. The advantages of the proposed RDRR-MUSIC algorithm are demonstrated using numerical simulation.

The contributions of this paper are as follows:

(1) We decouple DOA and range by reconstructing the steering vector, and all DOA information is fully used in PRF, which prevents DOA estimation performance degradation.
(2) We construct a transformation mechanism between DOA and range, and eliminate the error conduction caused by the extra decoupling operation, which prevents range estimation performance degradation.
(3) We achieve the automatic pairing of DOA estimates and range estimates.
(4) We reduce the large computational costs and realize rapid target localization using RDT and PRF.

The remainder of this paper can be summarized as follows: Section 2 shows the system model. The 2D-MUSIC SSF reconstruction, RDT, and PRF are presented in Section 3. Section 4 comprises the numerical simulations. The conclusion can be found in Section 5.

2. System Model

As shown in Figure 1, the transmitter and receiver of collocated FDA-MIMO radar contain M and N sensors, respectively, and they are all uniform linear arrays with $d = \lambda/2$ interspace, where λ is the wavelength. The radiated frequency of the mth sensor is defined as follows:

$$f_m = f_0 + (m-1)\Delta f, m = 1, 2, ..., M \tag{1}$$

where f_0 is the carrier frequency, Δf is the frequency offset between adjacent sensors, and $\Delta f \ll f_0$. We assume that K point targets need to be located, and (θ_k, r_k) are the DOA and range of the kth target, $k = 1, 2, ..., K$. Then, the output signal after matched filtering can be given as follows: [6]

$$\mathbf{x}(t) = \mathbf{A}\mathbf{s}(t) + \mathbf{n}(t) \tag{2}$$

where $\mathbf{s}(t) = [s_1(t), s_2(t), ..., s_K(t)]^T$ is the signal vector, $s_k(t) = \sigma_k e^{j2\pi f_{d_k}(t-\tau_0)} e^{-4\pi f_0 r_k/c}$, σ_k is the radar cross section, f_{d_k} is the Doppler frequency, $\tau_0 = 2r/c$ is the time delay between the first sensor and target, c is light speed, and $\mathbf{n}(t) \sim \mathcal{CN}(\mathbf{0}, \delta^2\mathbf{I})$ is the noise vector. Moreover, $\mathbf{A}$ is the steering matrix which is given as follows:

$$\begin{aligned} \mathbf{A} &= [\mathbf{a}_r(\theta_1) \otimes \mathbf{a}_t(\theta_1, r_1), \ldots, \mathbf{a}_r(\theta_K) \otimes \mathbf{a}_t(\theta_K, r_K)] \\ &= [\mathbf{a}(\theta_1, r_1), \ldots, \mathbf{a}(\theta_K, r_K)] \end{aligned} \tag{3}$$

where $\mathbf{a}(\theta_k, r_k)$ is the steering vector, $\otimes$ is the Kronecker product, and $\mathbf{a}_r(\theta_k)$ is the receive steering vector, which is defined as follows:

$$\mathbf{a}_r(\theta_k) = \left[1, e^{j2\pi d \sin(\theta_k)/\lambda}, \ldots, e^{j2\pi(N-1)d \sin(\theta_k)/\lambda}\right]^T \tag{4}$$

$\mathbf{a}_t(\theta_k, r_k)$ is the transmit steering vector, which is defined as follows:

$$\mathbf{a}_t(\theta_k, r_k) = \mathbf{a}_t(\theta_k) \oplus \mathbf{a}_t(r_k) \tag{5}$$

$\oplus$ is the Hadamard product and $\mathbf{a}_t(\theta_k)$ is the DOA-dependent transmit steering vector, defined as follows:

$$\mathbf{a}_t(\theta_k) = \left[1, e^{j2\pi d \sin(\theta_k)/\lambda}, \ldots, e^{j2\pi(M-1)d \sin(\theta_k)/\lambda}\right]^T \tag{6}$$

$\mathbf{a}_t(r_k)$ is the range-dependent transmit steering vector, defined as follows:

$$\mathbf{a}_t(r_k) = \left[1, e^{-j4\pi\Delta f r_k/c}, \ldots, e^{-j4\pi(M-1)\Delta f r_k/c}\right]^T \tag{7}$$

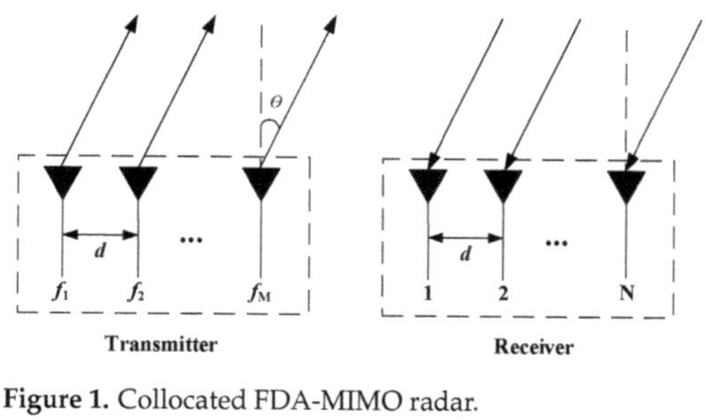

Figure 1. Collocated FDA-MIMO radar.

When we consider finite samples, the covariance matrix (CM) of the output signal can be presented as follows:

$$\hat{\mathbf{R}} = \frac{1}{L}\sum_{l=1}^{L} \mathbf{x}(l)\mathbf{x}^H(l) \tag{8}$$

where L is the snapshot.

3. Proposed Method

We conduct eigenvalue decomposition (CED) for $\hat{\mathbf{R}}$, resulting in the following:

$$\hat{\mathbf{R}} = \mathbf{E}_s\mathbf{D}_s\mathbf{E}_s^H + \mathbf{E}_n\mathbf{D}_n\mathbf{E}_n^H \tag{9}$$

where $\mathbf{E}_s \in \mathbb{C}^{NM\times K}$ is the signal subspace containing the eigenvectors corresponding to the largest K eigenvalues and $\mathbf{E}_n \in \mathbb{C}^{NM\times(NM-K)}$ is the noise subspace that consists of eigenvectors corresponding to the other eigenvalues [16]. We construct 2D-MUSIC SSF as in [15], as follows:

$$f_{2D-MUSIC}(\theta, r) = \frac{1}{\mathbf{a}^H(\theta, r)\mathbf{E}_n\mathbf{E}_n^H\mathbf{a}(\theta, r)} \tag{10}$$

We can find K spectrum peaks using 2D SPS, and the entries of the peaks denote the locations of the targets. However, the conventional 2D-MUSIC algorithm cannot be utilized directly in practice due to the heavy computational burden caused by 2D SPS. Therefore, we propose the RDRR-MUSIC algorithm to overcome this drawback.

We reconstruct the steering vector with the decoupling of DOA and range, as follows:

$$\begin{aligned}\mathbf{a}(\theta, r) &= \mathbf{a}_r(\theta) \otimes [diag(\mathbf{a}_t(\theta))\mathbf{a}_t(r)] \\ &= [\mathbf{a}_r(\theta) \otimes diag(\mathbf{a}_t(\theta))]\mathbf{a}_t(r) \\ &= \mathbf{D}(\theta)\mathbf{a}_t(r)\end{aligned} \tag{11}$$

where $diag(.)$ represents a diagonal matrix where the diagonal entries are composed of the entries of a vector. Note that $\mathbf{a}_r(\theta)$ and $\mathbf{a}_t(\theta)$ are combined into $\mathbf{D}(\theta)$, which is independent of the range. The 2D-MUSIC SSF can be reconstructed as follows:

$$f_{RC-MUSIC}(\theta, r) = \frac{1}{\mathbf{a}_t^H(r)\mathbf{D}^H(\theta)\mathbf{E}_n\mathbf{E}_n^H\mathbf{D}(\theta)\mathbf{a}_t(r)} \tag{12}$$

Then, we conduct RDT, to alleviate the computational complexity.

We construct the optimization function as in [15], as follows:

$$\min_{\theta, r} \mathbf{a}_t^H(r)\mathbf{Q}(\theta)\mathbf{a}_t(r) \quad \text{s.t. } \mathbf{e}_1^H\mathbf{a}_t(r) = 1 \tag{13}$$

where $\mathbf{Q}(\theta) = \mathbf{D}^H(\theta)\mathbf{E}_n\mathbf{E}_n^H\mathbf{D}(\theta)$, $\mathbf{e}_1 = [1, 0, ..., 0]^T \in \mathbb{R}^{M\times 1}$. Then, the loss function is as in [15], as follows:

$$L(\theta, r) = \mathbf{a}_t^H(r)\mathbf{Q}(\theta)\mathbf{a}_t(r) - \eta\left(\mathbf{e}_1^H\mathbf{a}_t(r) - 1\right) \tag{14}$$

where η is the regularization coefficient. Then, we can obtain the following:

$$\frac{\partial L(\theta, r)}{\partial \mathbf{a}_t(r)} = 2\mathbf{Q}(\theta)\mathbf{a}_t(r) - \eta\mathbf{e}_1 = 0 \tag{15}$$

We consider $\mathbf{e}_1^H\mathbf{a}_t(r) = 1$; therefore, we can obtain the following:

$$\mathbf{a}_t(r) = \frac{\mathbf{Q}^{-1}(\theta)\mathbf{e}_1}{\mathbf{e}_1^H\mathbf{Q}^{-1}(\theta)\mathbf{e}_1} \tag{16}$$

Note that the transformation mechanism between DOA and range is constructed, and the extra decoupling operation is undesired because $\mathbf{a}_t(r)$ is independent from DOA. The optimization function can be further reconstructed as follows:

$$\hat{\theta} = \min_{\theta} \frac{1}{\mathbf{e}_1^H \mathbf{Q}^{-1}(\theta)\mathbf{e}_1} = \min_{\theta} \frac{det(\mathbf{Q}(\theta))}{compan(\mathbf{Q}(\theta))_{1,1}} \tag{17}$$

where $det(.)$ is the determinant and $compan(.)$ is the companion matrix. We can estimate the DOA by searching the zero value of $det(\mathbf{Q}(\theta))$, and the 2D SPS is converted into 1D SPS using RDT. In order to further reduce the computational complexity, we construct the following symbol vector:

$$\begin{aligned} \mathbf{Q}(z) = {} & \left[z^{N-1}\mathbf{a}_r\left(z^{-1}\right) \otimes diag\left(z^{M-1}\mathbf{a}_t\left(z^{-1}\right)\right)\right]^T \mathbf{E}_n \\ & \times \mathbf{E}_n^H [\mathbf{a}_r(z) \otimes diag(\mathbf{a}_t(z))] \end{aligned} \tag{18}$$

where $z = e^{j2\pi d \sin(\theta)/\lambda}$ is the symbol variable, z^{N-1} and z^{M-1} are utilized to avoid negative power series of z, $\mathbf{a}_r(z) = \left[1, z, \ldots, z^{N-1}\right]^T$, and $\mathbf{a}_t(z) = \left[1, z, \ldots, z^{M-1}\right]^T$. We can find K roots $\hat{z}_1, \hat{z}_2, \ldots, \hat{z}_K$ by conducting PRF to $det(\mathbf{Q}(z)) = 0$, and the DOA estimates $\hat{\theta}_k$ can be given as follows:

$$\hat{\theta}_k = \arcsin[angle(\hat{z}_k)\lambda/2\pi d], \quad k = 1, 2, \ldots, K \tag{19}$$

where $angle(.)$ is the phase angle of a complex number.

Considering Equation (16), we obtain the following:

$$\hat{\mathbf{a}}_t(r_k) = \frac{\mathbf{Q}^{-1}(\hat{\theta}_k)\mathbf{e}_1}{\mathbf{e}_1^H \mathbf{Q}^{-1}(\hat{\theta}_k)\mathbf{e}_1} \tag{20}$$

We construct the least square fitting (LSF) of the range, as follows:

$$\hat{\mathbf{g}}_k = \min_{\mathbf{g}_k} \|\mathbf{G}\mathbf{g}_k - angle(\hat{\mathbf{a}}_t(r_k))\|^2, k = 1, 2, \ldots, K \tag{21}$$

where $\mathbf{G} = [\mathbf{1}_M, \mu]$, $\mu = [0, -4\pi\Delta f/c, \ldots, -4\pi(M-1)\Delta f/c]^T$, $\mathbf{1}_M = [1, 1, \ldots, 1]^T \in \mathbb{R}^{M\times 1}$, $\mathbf{g}_k = [\epsilon_k, r_k]^T$, and ϵ_k is residual. The range estimates $\hat{r}_k$ can be given as follows:

$$\hat{\mathbf{g}}_k = \left(\mathbf{G}^H\mathbf{G}\right)^{-1}\mathbf{G}^H angle(\hat{\mathbf{a}}_t(r_k)), k = 1, 2, \ldots, K \tag{22}$$

where $\hat{r}_k$ is the second element of $\hat{\mathbf{g}}_k$.

Remark 1. *The transformation mechanism between the DOA and the range is constructed as Equation (16), so the DOA estimates and range estimates are automatically paired.*

Remark 2. *The computational complexity of the proposed RDRR-MUSIC algorithm includes the following parts: the CM construction in Equation (8) requires $O((MN)^2L)$, the CED in Equation (9) needs $O((MN)^3)$, the symbol vector construction and PRF in Equation (18) requires $O(M^2(MN-K) + (2M(N-1))^3)$, and LSF in Equation (21) needs $O((3M+1)K)$, which avoids the heavy computational cost of SPS.*

Remark 3. *The proposed RDRR-MUSIC algorithm can also be utilized in the multipath interference scenario, but the rank of the covariance matrix in Equation (8) needs to be recovered via spatial smoothing [17].*

The specific steps of the proposed RDRR-MUSIC algorithm are concluded as follow Algorithm 1:

Algorithm 1 RDRR-MUSIC

Input: The output signal $\mathbf{x}(t)$;
1: Calculate **R** and **En** using Equations (8) and (9);
2: Reconstruct 2D-MUSIC SSF utilizing reconstructed steering vector through Equation (12);
3: Perform RDT and PRF to obtain DOA estimates using Equation (19);
4: Perform LSF to obtain range estimates using Equation (21);
Output: $(\hat{\theta}_k, \hat{r}_k), k = 1, 2, ..., K$

4. Complexity Analysis

The computational complexity of the proposed RDRR-MUSIC algorithm, the ESPRIT algorithm [11], the conventional 2D-MUSIC algorithm [14], and the RD-MUSIC algorithm [15] are described in Table 1, where $\Delta\theta$ and Δr are the DOA and range search scopes, and θ_d and r_d are the DOA and range search step widths, respectively. Meanwhile, the complexity comparison versus the number of receive sensors is presented in Figure 2, where $M = 6$, $L = 200$, $\Delta\theta = 180°$, $\Delta r = 500$ m,$\theta_d = 0.01°$, and $r_d = 0.01$ m. The complexity comparison versus the number of snapshots is shown in Figure 3, where $M = 6$, $N = 4$, $\Delta\theta = 180°$, $\Delta r = 500$ m, $\theta_d = 0.01°$, and $r_d = 0.01$ m. It is obvious that the computational complexity of the proposed RDRR-MUSIC is much lower than that of the conventional 2D-MUSIC algorithm and RD-MUSIC algorithm, and is almost the same as for the ESPRIT algorithm. The reason for this is that 2D SPS and 1D SPS are successfully eliminated by RDT and PRF.

Table 1. Computational complexity comparison of different algorithms.

Algorithm	Complexity
RDRR-MUSIC	$O((MN)^2L + (MN)^3 + M^2(MN-K) + (2M(N-1))^3 + (3M+1)K)$
ESPRIT	$O((MN)^2L + (MN)^3 + 2K^2M(N-1) + 2K^2(M-1)N + 6K^3)$
2D-MUSIC	$O((MN)^2L + (MN)^3 + (MN+1)(MN-K)\Delta\theta\Delta r/\theta_d r_d)$
RD-MUSIC	$O((MN)^2L + (MN)^3 + (M^2(N+1)(MN-K) + M^3)\Delta\theta/\theta_d)$

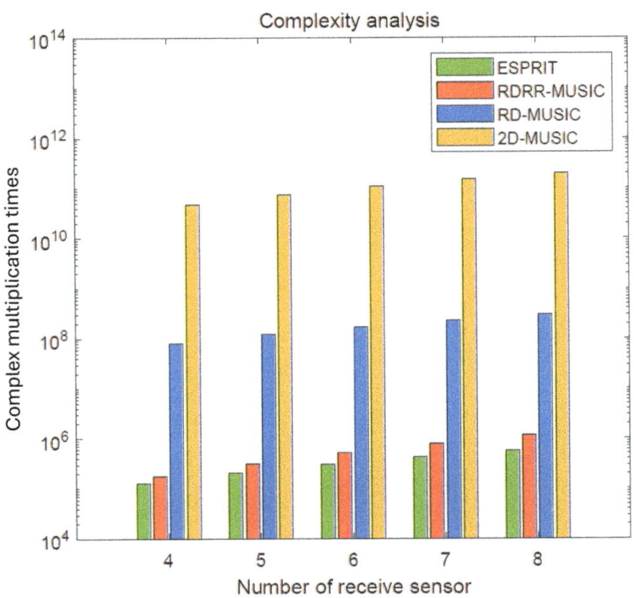

Figure 2. Complexity comparison versus the number of receive sensors.

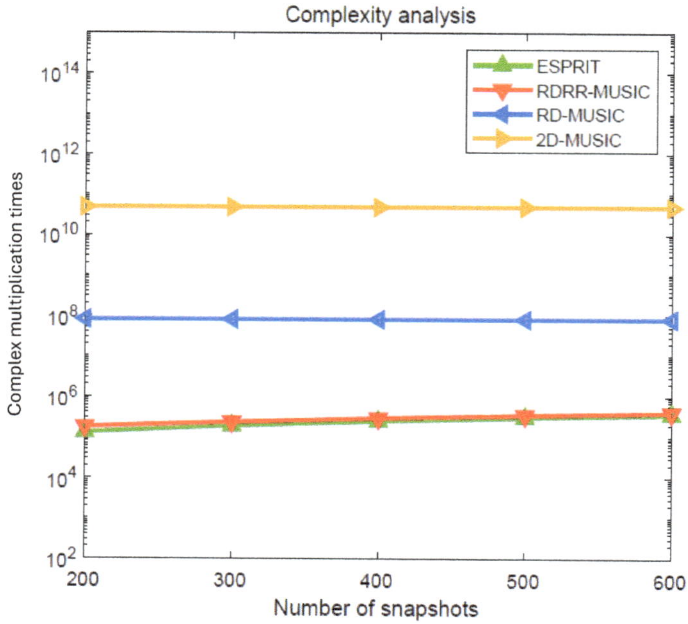

Figure 3. Complexity comparison versus the number of snapshots.

In order to further demonstrate the effectiveness of the proposed RDRR-MUSIC algorithm, a comparison of the actual computation time is shown in Table 2, where $M = 3$, $N = 3$, $L = 400$, $\Delta\theta = 180°$, $\Delta r = 500$ m, $\theta_d = 0.01°$, and $r_d = 0.01$ m. Different algorithms are computed by the MATLAB R2021b with Intel Core i5-12400 @2.5 GHz and 16GB RAM. We found that the computation time of the proposed RDRR-MUSIC algorithm is shorter than that of the conventional 2D-MUSIC algorithm [14] and the RD-MUSIC algorithm [15].

Table 2. Computation time comparison of different algorithms.

Algorithm	Complexity Multiplications	Computation Time, s
ESPRIT	3.3507×10^4	0.552×10^{-3}
RDRR-MUSIC	3.4941×10^4	0.4704×10^{-1}
RD-MUSIC	4.8931×10^6	0.8895×10^{-1}
2D-MUSIC	5.4×10^9	6.1463

5. Numerical Simulations

We perform numerical simulations to validate the superiority and effectiveness of the proposed RDRR-MUSIC algorithm. Different algorithms are compared with the proposed RDRR-MUSIC, including the ESPRIT algorithm [11], the conventional 2D-MUSIC algorithm [14], the RD-MUSIC algorithm [15], and the Cramér–Rao bound (CRB). The root mean square error (RMSE) is utilized to evaluate all the algorithms mentioned above and is defined as follows:

$$RMSE_\beta = \sqrt{\frac{1}{KP}\sum_{k=1}^{K}\sum_{p=1}^{P}\left(\hat{\beta}_{k,p} - \beta_k\right)^2} \tag{23}$$

where $\hat{\beta}_{k,p}$ is the DOA or range estimate of the kth target, which is conducted by the pth trail, β_k is the real DOA or range of the kth target, and P denotes the number of Monte Carlo trials. Table 3 shows the simulation parameter configuration, where $\Delta f \ll f_0$. Note that the maximum unambiguous estimation range is $r_{max} = c/2\Delta f$ [6].

Table 3. Parameter configuration.

Parameter	M	N	f_0	Δf	K	(θ_k, r_k)
Value	6	8	10 GHz	300 KHz	3	$(10°, 200)$ m $(20°, 220)$ m $(30°, 240)$ m

Example 1. *The estimates of the proposed RDRR-MUSIC algorithm are shown in Figure 4, where the signal-to-noise ratio (SNR) is $SNR = 5$ dB, $L = 400$, and $P = 200$. We found that the estimates of three targets are close to the corresponding real values, and the DOA estimates and range estimates are automatically paired, which verifies the reliability of the proposed RDRR-MUSIC algorithm.*

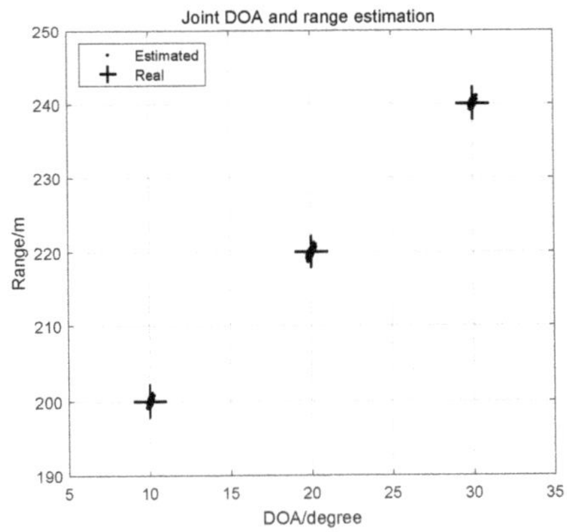

Figure 4. Scatter figure.

Example 2. *The performance comparison of different algorithms versus SNR are presented in Figures 5 and 6, where $L = 400$, $P = 600, \theta_d = 0.01°$, and $r_d = 0.01$ m. In Figures 5 and 6, we can observe that the RMSE curves of the proposed RDRR-MUSIC algorithm, the conventional 2D-MUSIC algorithm, and the RD-MUSIC algorithm almost overlap in $SNR > 0dB$ cases, which means that they enjoy identical estimation accuracy. In $SNR \leq 0$ dB cases, the proposed RDRR-MUSIC algorithm enjoys a better estimation performance than the RD-MUSIC algorithm, because the proposed RDRR-MUSIC algorithm fully exploits all DOA information when conducting PRF, and the error conduction caused by the extra decoupling operation is eliminated. Moreover, the proposed RDRR-MUSIC algorithm outperforms the conventional 2D-MUSIC algorithm in $SNR \leq 0$ dB cases. The reason for this is that, because the estimates of the conventional 2D-MUSIC algorithm are generated by 2D SPS, the grid mismatch problem cannot be solved, even in a dense search grid. Meanwhile, the DOA searching error and range searching error will interfere with each other, deteriorating the 2D searching accuracy. However, the estimates of the proposed RDRR-MUSIC algorithm are calculated using PRF, which avoids the grid mismatch problem.*

The performance comparison of different algorithms versus the number of snapshots are also plotted in Figures 7 and 8, where $SNR = 10$ dB, $P = 600$, $\theta_d = 0.01°$, and $r_d = 0.01$

m. In Figures 7 and 8, the proposed RDRR-MUSIC algorithm, the conventional 2D-MUSIC algorithm, and the RD-MUSIC algorithm have identical estimation performances, and they all outperform the ESPRIT algorithm. However, the computational complexity of the proposed RDRR-MUSIC algorithm is much lower than that of the conventional 2D-MUSIC algorithm and the RD-MUSIC algorithm, which means that the DOA and range can be rapidly estimated without performance degradation. This simulation demonstrates the superiority of the proposed RDRR-MUSIC algorithm.

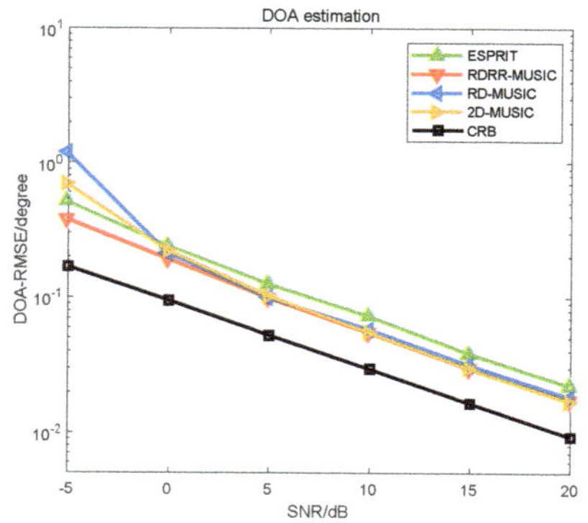

Figure 5. DOA estimation performance comparison of different algorithms versus SNR.

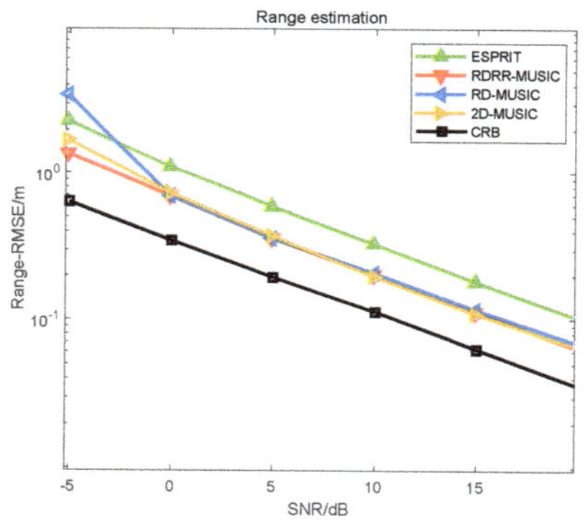

Figure 6. Range estimation performance comparison of different algorithms versus SNR.

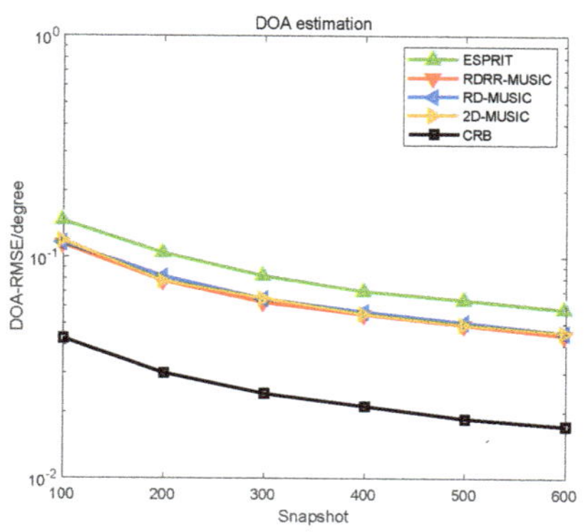

Figure 7. DOA estimation performance comparison of different algorithms versus the number of snapshots.

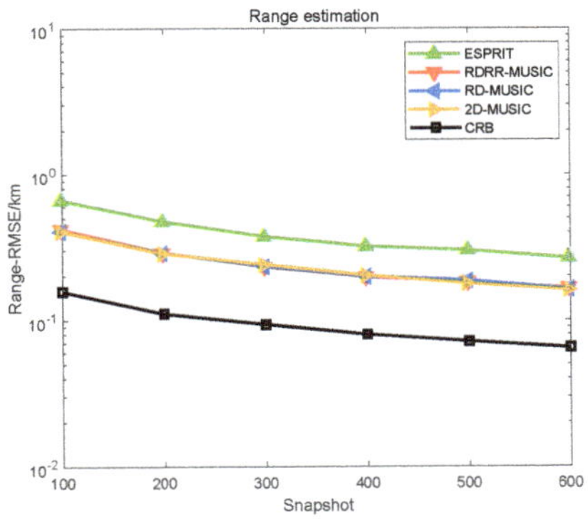

Figure 8. Range estimation performance comparison of different algorithms versus the number of snapshots.

6. Conclusions

The RDRR-MUSIC algorithm, used to rapidly locate targets using the FDA-MIMO radar, was proposed in this paper. We decoupled the DOA and range by reconstructing the steering vector, and achieved the automatic pairing of DOA estimates and range estimates. Meanwhile, we realized rapid target localization, but without performance degradation. Numerical simulations were used to demonstrate the effectiveness and superiority of the proposed RDRR-MUSIC algorithm. However, the accuracy of the DOA and range estimates were limited by the system's framework, and the array aperture and signal bandwidth need to be expanded. In future research, we will propose sparse frameworks and the corresponding algorithms to further improve the DOA and range estimation performance.

Author Contributions: Conceptualization, C.W.; methodology, C.W. and Z.Z.; software, C.W.; validation, C.W. and W.-Q.W.; formal analysis, C.W.; investigation, C.W. and Z.Z.; resources, W.-Q.W.; data curation, C.W.; writing—original draft preparation, C.W.; writing—review and editing, Z.Z.; visualization, C.W.; supervision, W.-Q.W.; project administration, C.W.; funding acquisition, C.W. and Z.Z. All authors have read and agreed to the published version of the manuscript.

Funding: This work was supported in part by the National Natural Science Foundation of China (No. 62301111, 62171089), and in part by the Science and Technology Plan Project of Huzhou City (No. 2022GZ03).

Institutional Review Board Statement: Not applicable.

Informed Consent Statement: Not applicable.

Data Availability Statement: The data used to support the findings of this study are available from the corresponding author upon request.

Acknowledgments: We sincerely thank all the reviewers for their patient reviews of and valuable suggestions for this article.

Conflicts of Interest: We declare that we have no financial and personal relationships with other people or organizations that can inappropriately influence our work, there is no professional or other personal interest of any nature or kind in any product, service, and/or company that could be construed as influencing the position presented in, or the review of, the manuscript entitled.

References

1. Li, J.; Li, P.; Li, P.; Tang, L.; Zhang, X.; Wu, Q. Self-Position Awareness Based on Cascade Direct Localization Over Multiple Source Data. *IEEE Trans. Intell. Transp. Syst.* **2024**, *25*, 796–804. [CrossRef]
2. Lai, X.; Zhang, X.; Liu, W.; Li, J. Sparse Enhancement of MIMO Radar Exploiting Moving Transmit and Receive Arrays for DOA Estimation: From the Perspective of Synthetic Coarray. *IEEE Trans. Signal Process.* **2024**, *72*, 4022–4036. [CrossRef]
3. Antonik, P.; Wicks, M.C.; Griffiths, H.D.; Baker, C.J. Frequency diverse array radars. In Proceedings of the 2006 IEEE Conference on Radar, Verona, NY, USA, 24–27 April 2006 . [CrossRef]
4. Zubair, M.; Ahmed, S.; Alouini, M.S. Frequency Diverse Array Radar: New Results and Discrete Fourier Transform Based Beampattern. *IEEE Trans. Signal Process.* **2020**, *68*, 2670–2681. [CrossRef]
5. Tan, M.; Wang, C.; Xue, B.; Xu, J. A Novel Deceptive Jamming Approach Against Frequency Diverse Array Radar. *IEEE Sensors J.* **2021**, *21*, 8323–8332. [CrossRef]
6. Xu, J.; Liao, G.; Zhu, S.; Huang, L.; So, H.C. Joint Range and Angle Estimation Using MIMO Radar With Frequency Diverse Array. *IEEE Trans. Signal Process.* **2015**, *63*, 3396–3410. [CrossRef]
7. Wang, C.; Zhang, X.; Li, J. FDA-MIMO radar for 3D localization: Virtual coprime planar array with unfolded coprime frequency offset framework and TRD-MUSIC algorithm. *Digit. Signal Process.* **2021**, *113*, 1–15. [CrossRef]
8. Wang, C.; Zhang, X.; Li, J. FDA-MIMO radar for DOD, DOA, and range estimation: SA-MCFO framework and RDMD algorithm. *Signal Process.* **2021**, *188*, 1–11. [CrossRef]
9. Wang, W.; Shao, H. Range-Angle Localization of Targets by A Double-Pulse Frequency Diverse Array Radar. *IEEE J. Sel. Top. Signal Process.* **2014**, *8*, 106–114. [CrossRef]
10. Wang, W.; So, H.C. Transmit Subaperturing for Range and Angle Estimation in Frequency Diverse Array Radar. *IEEE Trans. Signal Process.* **2014**, *62*, 2000–2011. [CrossRef]
11. Li, B.; Bai, W.; Zheng, G. Successive ESPRIT algorithm for joint DOA-range-polarization estimation with polarization sensitive FDA-MIMO radar. *IEEE Access* **2018**, *6*, 36376–36382. [CrossRef]
12. Cui, C.; Xu, J.; Gui, R.; Wang, W.; Wu, W. Search-Free DOD, DOA and Range Estimation for Bistatic FDA-MIMO Radar. *IEEE Access* **2018**, *6*, 15431–15445. [CrossRef]
13. Chen, H.; Wang, W.; Liu, W. Augmented Quaternion ESPRIT-Type DOA Estimation With a Crossed-Dipole Array. *IEEE Commun. Lett.* **2020**, *24*, 548–552. [CrossRef]
14. Jie, X.; Wang, W.; Gao, K. FDA-MIMO radar Range–Angle estimation: CRLB, MSE, and resolution analysis. *IEEE Trans. Aerosp. Electron. Syst.* **2018**, *54*, 284–294.
15. Zhang, X.; Xu, L.; Xu, L.; Xu, D. Direction of Departure (DOD) and Direction of Arrival (DOA) Estimation in MIMO Radar with Reduced-Dimension MUSIC. *IEEE Commun. Lett.* **2010**, *14*, 1161–1163. [CrossRef]
16. Schmidt, R. Multiple emitter location and signal parameter estimation. *IEEE Trans. Antennas Propag.* **1986**, *34*, 276–280. [CrossRef]
17. Pan, J.; Sun, M.; Wang, Y.; Zhang, X.; Li, J.; Jin, B. Simplified Spatial Smoothing for DOA Estimation of Coherent Signals. *IEEE Trans. Circuits Syst. II Express Briefs* **2023**, *70*, 841–845. [CrossRef]

sensors

MDPI

Communication

Efficient Aperture Fill Time Correction for Wideband Sparse Array Using Improved Variable Fractional Delay Filters

Jie Gu [1,*], Min Xu [1], Wenjing Zhou [2] and Mingwei Shen [2]

[1] National Key Laboratory of Electromagnetic Space Security, Chengdu 610036, China; minxumin@163.com
[2] College of Information Science and Engineering, Hohai University, Nanjing 210098, China; 15637357907@163.com (W.Z.); smw_nuaa@hhu.edu.cn (M.S.)
* Correspondence: cetc_gujie@163.com; Tel.: +86-18215606483

Abstract: To solve the problem of aperture fill time (AFT) for wideband sparse arrays, variable fractional delay (VFD) FIR filters are applied to eliminate linear coupling between spatial and time domains. However, the large dimensions of the filter coefficient matrix result in high system complexity. To alleviate the computational burden of solving VFD filter coefficients, a novel multi–regultion minimax (MRMM) model utilizing the sparse representation technique has been presented. The error function is constrained by the introduction of L2–norm and L1–norm regularizations within the minimax criterion. The L2–norm effectively resolves the problems of overfitting and non–unique solutions that arise in the sparse optimization of traditional minimax (MM) models. Meanwhile, the use of multiple L1–norms enables the optimal design of the smallest sub–filter number and order of the VFD filter. To solve the established nonconvex model, an improved sequential–alternating direction method of multipliers (S–ADMM) algorithm for filter coefficients is proposed, which utilizes sequential alternation to iteratively update multiple soft–thresholding problems. The experimental results show that the optimized VFD filter reduces system complexity significantly and corrects AFT effectively in a wideband sparse array.

Keywords: wideband sparse array; variable fractional delay (VFD); minimax criterion; ADMM algorithm

1. Introduction

A wideband sparse array has the advantages of improving spatial resolution, reducing hardware complexity, and improving target detection and recognition performance. However, the aperture fill time (AFT) phenomenon, inherent to wideband arrays, significantly diminishes the performance of beamforming. Variable fractional delay (VFD) filters involve the implementation of precise variable delay in digital systems and are widely used for aperture correction in wideband arrays [1,2]. High–precision delay filters are usually represented as a sum of integer and fractional terms. While cascaded registers allow for the implementation of integer terms, fractional terms require more sophisticated techniques [3,4]. The pioneering Farrow structure [5], known for its flexible and efficient delay correction, modifies and extends the traditional structure, effectively reducing implementation complexity and improving design accuracy.

In general, the design of a Farrow structure VFD (FS–VFD) filter primarily involves the minimax method [6,7] and the weighted least squares (WLS) method [8,9]. The minimax method is implemented by minimizing the maximum magnitude value between the actual frequency response and the ideal frequency response, ensuring accuracy within the approximation error range [7]. Some studies have transformed the solution problem into convex optimization problems in order to ensure the optimal value of the coefficient solution; it still requires numerous iterations and updates, leading to high computational complexity. The WLS design is implemented by minimizing the mean square error of the two frequencies, yielding an optimal closed solution [8]. However, the higher–order filter design leads to a cascading increase in computational complexity for both design criterion, necessitating additional storage resources [9]. High computational complexity and unstable values affect the efficiency of FS–VFD filters, so an ideal design approach needs to ensure accuracy while reducing computational complexity.

FS–VFD FIR filters comprise several sub–filters with a large number of coefficients, distinguishing them from the numerical scalars in digital FIR filters [10]. To reduce computation, the literature [11] reduces the number of filter coefficients using the WLS design criterion and the (inverse) symmetry of the coefficients but does not take into account the change in system complexity. Filter coefficient sparse processing is an optimization technique aimed at reducing system complexity and improving processing efficiency by reducing the number of non–zero coefficients in a filter [12,13]. Through sparse optimization, we can obtain a more compact and efficient filter structure, which is especially important for resource–constrained environments or application scenarios requiring real–time performance. The literature [14] studies VFDs with sparse coefficients for reducing computation. Leveraging the polynomial properties of the coefficients, a two–stage technique is proposed for a VFD FIR filter, aimed at obtaining a least squares solution under sparse target coefficients. To reduce system complexity and computation of the Farrow structure, the literature [15] considers sparse–optimized Farrow structure VFD (SFS–VFD) filters based on least squares and coefficient (inverse) symmetry to ensure high accuracy and efficient correction of aperture effects. It is noteworthy that while much research focuses on implementing FS–VFD filters with the WLS criterion, there remains a gap in related research based on the minimax criterion [16].

Therefore, this paper proposes a new multi–regulation minimax sparse model for VFD design. The error function is constrained by introducing $L2$–norm and $L1$–norm regularizations under the minimax criterion. The $L2$–norm solves the overfitting and non–unique solution problems of the traditional MM model in sparse optimization, and the $L1$–norm achieves the optimal design of the smallest sub–filter number and order for VFD filters. An improved sequential ADMM algorithm is proposed for solving this nonconvex model. Subsequently, the proposed MRMM–VFD filters are applied to a wideband sparse array to correct the AFT. Finally, simulation results verify the effectiveness and feasibility of MRMM–VFD filters.

2. Problem Formulation

The ideal frequency response of a VFD filter is expressed as follows:

$$H_d(\omega, p) = e^{-j\omega p} \tag{1}$$

where $\omega \in [0, \alpha\pi]$ is the passband range, and $0 < \alpha \leq 1$, $p \in [-0.5, 0.5]$ is the fractional delay parameter range. The block diagram of the VFD filter structure based on the Farrow structure is shown in Figure 1. Therefore, the actual filter frequency response can be expressed as follows:

$$H(\omega, p) = \sum_{n=-N}^{N+1} \sum_{m=0}^{M} a(n, m) p^m e^{-j\omega n} \tag{2}$$

where $a(n, m)$ are the filter coefficients; M is the number of sub–filters; and N is the sub–filter order. Since VFD filters have coefficient (anti)symmetry, taking an odd number N as an example, the coefficients of the even sub–filter (with symmetry) and the odd sub–filter (with anti–symmetry) are split to obtain the following form:

$$H(\omega, p) = \sum_{m=0}^{M_e} c_m^T b_{em} p^{2m} - j \sum_{m=1}^{M_o} s_m^T b_{om} p^{2m-1} \tag{3}$$

where

$$\begin{cases} \mathbf{c}_{\mathrm{m}}^T = [\cos(\frac{1}{2}\omega)\cos(\frac{3}{2}\omega)\ldots\cos((N+\frac{1}{2})\omega)]^t \\ \mathbf{s}_m^T = [\sin(\frac{1}{2}\omega)\sin(\frac{3}{2}\omega)\ldots\sin((N+\frac{1}{2})\omega)]^t \end{cases} \tag{4}$$

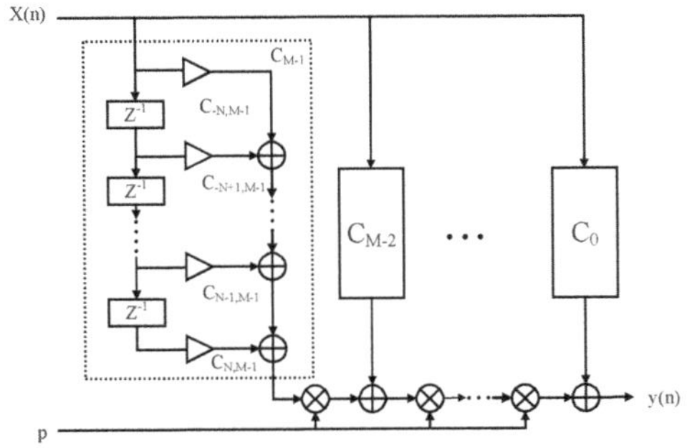

Figure 1. The Farrow structure the VFD FIR filter.

Convert it into the following form:

$$H(\omega,p) = f^T b_e - jg^T b_o \tag{5}$$

where

$$f_m^T = c_m^T p^{2m}, g_m^T = s_m^T p^{2m-1} \tag{6}$$

$$\begin{cases} f^T = \left[f_0^T f_1^T \cdots f_I^T \right] \\ g^T = \left[g_1^T g_2^T \cdots g_{I+1}^T \right] \end{cases} \tag{7}$$

The minimax design of a VFD filter refers to minimizing the peak error between the ideal frequency response and the actual frequency response. The solution problem for the filter coefficients of the minimax model is given as follows:

$$\text{minimize } \max|H(\omega,p) - H_d(\omega,p)| \tag{8}$$

The error function is defined as follows:

$$e_H(\omega,p) = H(\omega,p) - H_d(\omega,p) = e_R(\omega,p) - je_I(\omega,p) \tag{9}$$

where

$$\begin{cases} e_R(\omega,p) = \boldsymbol{f}\boldsymbol{b}_e - \cos(\omega p) \\ e_I(\omega,p) = \boldsymbol{g}\boldsymbol{b}_o - \sin(\omega p) \end{cases} \tag{10}$$

Then, problem (8) can be transformed into the following form:

$$\text{minimize } \|\boldsymbol{f}\boldsymbol{b}_e - \cos(\omega p) + j(\boldsymbol{g}\boldsymbol{b}_o - \sin(\omega p))\|_\infty \tag{11}$$

where $||\cdot||_\infty$ denotes the L_∞ norm.

3. Improved Sparse–Constrained Design for the VFD FIR Filter

3.1. The Cost Function Based on the Multi–Regulation Minimax Criterion

Building upon the traditional MM model established in the above section, SDP and SOCP algorithms are usually used to optimize the solution [17] but can easily lead to high system complexity. Owing to the characteristics of VFD filters with more near–zero coefficients, sparse constraint theory is introduced into the optimization of filter weight coefficients, and an improved MM model, the multi–regulation minimax (MRMM) model, is proposed in this paper. The model is introduced with $L2$ and $L1$ regularization to constrain the model coefficients. The $L2$–norm solves the overfitting and non-unique solution problems of the traditional MM model in sparse optimization [18]. The $L1$–norm realizes the design of the minimum number of sub–filters and orders of the VFD filter. The MRMM model is formulated as follows:

$$\text{minimize}||\Phi x-\psi||_\infty+||y||_2^2+\lambda_1\sum_{n=0}^{(N+1)(M+1)-1}\left\|\sum_{k=0}^{n}x_{(N+1)(M+1)-k}\right\|_1+\lambda_2\sum_{n=1}^{N+1}\sum_{m=0}^{M}\left\|\sum_{k=1}^{n}x_{(N+1-k)+M(N+1)}\right\|_1 \quad (12)$$

where the filter coefficient vector $\boldsymbol{x}=[\boldsymbol{b_e},\boldsymbol{b_o}]^T$ and the parameters Φ and ψ in the term $||\cdot||_\infty$ are related to Equation (11) in Section 2. According to reference [19], the diagonal elements of H are associated with the symmetry of the coefficients of the filter. y is the vector associated with the real impulse response $a(n,m)$ of the filter. According to coefficient symmetry $b(n,m)=2a(n,m)$, for $m=0,1,\cdots,M$, we define that the diagonal elements of H are $h_1=h_2=\ldots=h_{(N+1)(M+1)}=1/2$, and the diagonal matrix H can be reduced to a multiple of the unit matrix $I_{(N+1)(M+1)}$. In the L_∞ term, we give

$$\Phi=\begin{bmatrix}f & 0\\ 0 & g\end{bmatrix},\ \psi=\begin{bmatrix}\cos(\omega p)\\ \sin(\omega p)\end{bmatrix} \quad (13)$$

The first and second terms in problem (12) denote the minimax error of the VFD filter; the third term denotes the optimal filter order constraint; and the fourth term denotes the optimal number of sub–filters constraint. The last two sparse terms should essentially be formulated as $L0$–norm problems. However, they cannot be solved directly due to NP Hard and a highly nonconvex nature. Therefore, the last two sparse terms are converted to $L1$–norm form after relaxation [20,21]. It can be seen that the orders of all sub–filters are optimized simultaneously by the second $L1$–norm term, which has a complex nested summation process and cannot be solved directly. Based on the properties of paradigm triangular inequality $||\sum_i b_i||\leq\sum_i||b_i||$ [22], problem (12) can be approximated as follows:

$$\text{minimize}||\Phi x-\psi||_\infty+x^THx+\lambda_1\sum_{n=0}^{(N+1)(M+1)-1}\sum_{k=0}^{n}\left|x_{(N+1)(M+1)-k}\right|+\lambda_2\sum_{n=1}^{N+1}\sum_{k=1}^{n}\left[\left|x_{(N+1-k)}\right|+\ldots+\left|x_{((N+1-k)+M(N+1))}\right|\right] \quad (14)$$

A further derivation gives the following:

$$\text{minimize}||\Phi x-\psi||_\infty+x^THx+\lambda_1\sum_{n=0}^{(N+1)(M+1)-1}((N+1)(M+1)-n)\left|x_{(N+1)(M+1)-n}\right|+\lambda_2\sum_{n=1}^{(M+1)(N+1)}rem[\frac{n}{(N+1)}]|x_n| \quad (15)$$

The last two items are integrated and reformulated in $L1$–norm form as follows:

$$\text{minimize}||\Phi x-\psi||_\infty+x^THx+\lambda_1||Qx||_1+\lambda_2||Sx||_1 \quad (16)$$

Q and S in the formula are, respectively, as follows:

$$\begin{cases}Q=diag(1,2,\ldots,(N+1)(M+1))\\ s_i=diag(1,2,\ldots,N+1)\\ S=diag(s_1,s_2,\ldots,s_{M+1})\end{cases} \quad (17)$$

3.2. S–ADMM Algorithm for Computing VFD Filter Coefficients

Since problem (16) needs to solve the optimization problem of the minimum sum of different paradigms, and the solution variables are multi–dimensional and complex, it is difficult for general optimization algorithms or tools to find the optimal solution. The ADMM algorithm combines the decomposability of the dual ascent method with the superior convergence of the multiplier method, making it suitable for solving large–scale distributed optimization problems.

The core idea of the ADMM algorithm is to decompose the original high–dimensional optimization problem into several low–dimensional subproblems, and gradually approach the optimal solution by solving these subproblems with alternating iterations. This decomposition and iteration strategy enables the algorithm to effectively deal with high–dimensional data and has better convergence and solving efficiency. However, the standard ADMM algorithm is only applicable to the problem of minimizing the sum of two norms. Therefore, in order to solve problem (3), the ADMM algorithm was improved and a sequential ADMM algorithm is proposed.

By introducing the auxiliary variables z_1, z_2, z_3, problem (16) is expressed as a minimum optimization problem with constraints as follows:

$$\begin{array}{l} \text{minimize } \|z_1 - \psi\|_\infty + \frac{1}{2}x^T Hx + \lambda_1\|z_2\|_1 + \lambda_2\|z_3\|_1 \\ \textit{subject to} \begin{bmatrix} \Phi \\ Q \\ S \end{bmatrix} x + \begin{bmatrix} -I \\ 0 \\ 0 \end{bmatrix} z_1 + \begin{bmatrix} 0 \\ -I \\ 0 \end{bmatrix} z_2 + \begin{bmatrix} 0 \\ 0 \\ -I \end{bmatrix} z_3 = 0 \end{array} \tag{18}$$

The variable x of the original problem is decomposed by the S–ADMM algorithm into multiple variables for sequential alternating iterations. Through constructing the augmented Lagrange function and updating the independent variables and the Lagrange multipliers, the iterations are repeated and stopped until the residuals reach a predetermined accuracy, thereby determining the optimal values of the variables. By defining the Lagrange multiplier vector $\boldsymbol{u} = [u_1, u_2, u_3]$, the following iterative solution formula is obtained using the dual gradient descent method:

$$\begin{cases} \overline{x}^{k+1} = \underset{x}{\operatorname{argmin}} L_\rho(x, z_1^k, z_2^k, z_3^k, u^k) \\ = \underset{x}{\operatorname{argmin}} \frac{1}{2}x^T Hx + \frac{\rho_1}{2}\|\Phi x - \psi - z_1^k + u_1^k\|_2^2 + \frac{\rho_2}{2}\|Qx - z_2^k + u_2^k\|_2^2 + \frac{\rho_3}{2}\|Sx - z_3^k + u_3^k\|_2^2 \\ x^{k+1} = \beta\overline{x}^k + (1 - \beta x^k) \\ z_1^{k+1} = \underset{z_1}{\operatorname{argmin}} L_\rho(x^k, z_1, u_1^k) = \underset{z_1}{\operatorname{argmin}} \|z_1 - \psi\|_\infty + \frac{\rho_1}{2}\|\Phi x^k - z_1 + u_1^k\|_2^2 \\ z_2^{k+1} = \underset{z_2}{\operatorname{argmin}} L_\rho(x^k, z_2, u_2^k) = \underset{x}{\operatorname{argmin}}\, \lambda_1\|z_2\|_1 + \frac{\rho_2}{2}\|Qx^k - z_2 + u_2^k\|_2^2 \\ z_3^{k+1} = \underset{z_3}{\operatorname{argmin}} L_\rho(x^k, z_3, u_3^k) = \underset{x}{\operatorname{argmin}}\, \lambda_2\|z_3\|_1 + \frac{\rho_3}{2}\|Sx^k - z_3 + u_3^k\|_2^2 \end{cases} \tag{19}$$

where β takes the value in the range of (0, 1]. It is the core correction parameter that ensures the convergence of the algorithm [23]. It is not difficult to find that solving for x is a quadratic optimization problem, while solving for the auxiliary variable z_1, z_2, z_3 is a multi–soft–threshold optimization problem. Multiple variables are iteratively updated in an alternating sequence and constraints are taken into account using the Lagrange multiplier u. The iterative formula for the Lagrange multiplier vector u is as follows:

$$\begin{cases} u_1^{k+1} = u_1^k + \rho_1(\Phi x^k - z_1^{k+1}) \\ u_2^{k+1} = u_2^k + \rho_2(Qx^k - z_2^{k+1}) \\ u_3^{k+1} = u_3^k + \rho_3(Sx^k - z_3^{k+1}) \end{cases} \tag{20}$$

The algorithm sets a predefined termination condition $\frac{\|x_{k+1} - x_k\|_2}{(N+1)(M+1)} \le \mu$ and determines whether the convergence criterion is met, where μ is the threshold value. If it is satisfied, the iteration is stopped; the optimal solution is found; and the final coefficients are obtained by using the symmetry of the coefficients; otherwise, the iteration is continued.

The coefficient vector x, i.e., the corresponding even sub–filter coefficients $\boldsymbol{b}_e$ and odd sub–filter coefficients $\boldsymbol{b}_o$, is computed according to the S–ADMM algorithm. The complete filter coefficient matrix is obtained using the following coefficient (inverse) symmetry formula:

$$b(n, m) = 2a(n, m), \text{ for } m = 0, 1, \cdots, M \tag{21}$$

4. Wideband Sparse Array AFT Correction Using MRMM–VFD

Sparse array design strives to optimally deploy sensors to achieve desirable beamforming characteristics, reduce system hardware costs, and lower computational complexity. In wideband signal models, sparse arrays can avoid conflicts caused by frequency spread of the signal. Reference [24] frames the design problem as optimally selecting P sensors out of N possible equally spaced locations ($P << N$), and a wideband sparse array design method based on the tapped delay line (TDL) structure was proposed.

Assume that all of the sensors are omnidirectional with the same response, and the signals impinge upon the array from the far field. The beamformer output $y[n]$ is a linear combination of differently delayed versions of the received array signals $x_m[n]$, $m = 0, \cdots, M-1$. The distance from the zeroth sensor to the subsequent sensor is denoted by d_m for $m = 0, \cdots, M-1$, with $d_0 = 0$. And the incident signal arrives at an angle θ.

The steering vector of the array as a function of the normalized frequency $\Omega = wT_s$ and the arrival angle θ is as follows:

$$\text{minimize}\|\Phi x - \psi\|_\infty + x^T H x + \lambda_1 \sum_{n=0}^{(N+1)(M+1)-1} \sum_{k=0}^{n} \left| x_{(N+1)(M+1)-k} \right| + \lambda_2 \sum_{n=1}^{N+1} \sum_{k=1}^{n} \left[\left| x_{(N+1-k)} \right| + \ldots + \left| x_{((N+1-k)+M(N+1))} \right| \right] \quad (22)$$

where $\mu_m = \frac{d_m}{cT_s}$ for $m = 0, 1, \cdots, M-1$, and $\{\cdot\}^T$ indicates a transpose operation. The response of the array is then given by the following:

$$P(\Omega, \theta) = \mathbf{w}^H \mathbf{s}(\Omega, \theta) \quad (23)$$

where $\mathbf{w}^H$ is the Hermitian transpose of the weight vector of the array, which is given by the following:

$$\mathbf{w} = \begin{bmatrix} \mathbf{w}_0^T & \mathbf{w}_1^T & \ldots & \mathbf{w}_{M-1}^T \end{bmatrix}^T \quad (24)$$

$$\mathbf{w}_m = \begin{bmatrix} \mathbf{w}_{m,0} & \mathbf{w}_{m,1} & \ldots & \mathbf{w}_{m,J-1} \end{bmatrix}^T \quad (25)$$

Then, a compressive sensing sparse algorithm is used to match the array response with the desired/reference response, resulting in a sparse wideband array distribution [24]. For the obtained wideband sparse array, the MRMM–VFD filter is employed for aperture crossing correction, with the specific operational diagram shown in Figure 2.

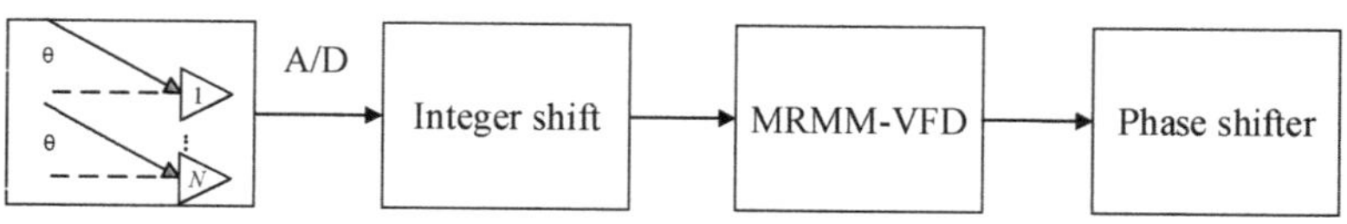

Figure 2. Aperture crossing correction of wideband sparse array.

The MRMM–VFD filter proposed in this paper has coefficient invariance, indicating that only the delay input needs to be altered while the filter coefficients remain unchanged when processing with different time delay parameters. In the aperture fill time correction of wideband sparse arrays, the MRMM–VFD greatly reduces computational complexity and system complexity.

5. Experimental Analysis

To verify the effectiveness of the designed MRMM sparse algorithm, the error results for different parameter settings are given in this section. Regarding the filter order N, a higher value results in a steeper frequency response of the filter, allowing for more accurate delay performance. However, excessively high orders may induce system instability, leading to increased noise and distortion; here, the value of N was chosen to be 30. The number of sub–filters M was set to 5, 7, 9, and 11. Regarding the value of the number of sub–filters, the amplitude response error decreases with an increase in the number of sub–filters. However, when it increases to a certain amount, the error reduction is no longer obvious. Additionally, an excessively large number of sub–filters can lead to unnecessary group phase delay errors [11,25]. The cutoff frequency parameter $\alpha = 1$, i.e., the full–pass bandwidth range was chosen for the experiment. In the MRMM algorithm, the penalty factors λ_1 and λ_2 are two important regularization parameters, which control the trade–off between data fitting and sparsity. We kept adjusting these two parameters with a sampling interval of 0.01, and the designed MRMM–VFD can achieve better performance when $\lambda_1 = 0.11$ and $\lambda_2 = 0.27$. To evaluate the error accuracy of the filter, the maximum magnitude error, the normalized mean square error (NMSE), and the maximum group delay error were used as evaluation metrics [25], and the expressions are shown, respectively, as follows:

$$\varepsilon_1 = \max\{20\log|H(\omega, p) - H_d(\omega, p)|\} \tag{26}$$

$$\varepsilon_2 = \left(\frac{\int_0^{\alpha\pi} \int_{-0.5}^{0.5} |H(\omega, p) - H_d(\omega, p)|_2 dp d\omega}{\int_0^{\alpha\pi} \int_{-0.5}^{0.5} |H_d(\omega, p)|_2 dp d\omega} \right)^{\frac{1}{2}} \tag{27}$$

$$\varepsilon_p = \max\{|\tau(\omega, p) - p|\} \tag{28}$$

Firstly, Figure 3a shows the convergence of primal and pairwise residuals in the S–ADMM algorithm, where the dotted lines are the set threshold; both of which can converge with an accuracy of 10^{-5}. The algorithm sets the maximum number of iterations to 100 but as can be seen in Figure 3b, the minimum of function (16) is actually found after 20 iterations, with a clear and fast trend. The results of the standard MM algorithm and the weighted least squares sparse algorithm are compared to demonstrate the effectiveness of the MRMM sparse model in this paper, and the error results of different algorithms are given in Table 1. When $M = 5$, the maximum magnitude error and the normalized mean square error of the MRMM sparse algorithm are larger than the standard MM algorithm, but as the number of sub–filters increases, the maximum magnitude error gradually approaches the standard MM algorithm. The normalized mean square error of the MRMM algorithm is smaller compared to the S–WLS sparse algorithm for different numbers of sub–filters. In addition, the more the number of sub–filters, the smaller the filter error, but too many sub–filters will make the system redundant, so $M = 9$ is chosen for practical applications. The amplitude–frequency response error plots of the VFD filters using the MRMM algorithm, the standard MM algorithm, and the S–WLS sparse algorithm for the $N = 30$, $M = 9$ condition are given as shown in Figure 4. In comparison, the MRMM algorithm is flatter than the S–WLS sparse algorithm. The maximum magnitude error is −19.81 dB for the MRMM algorithm and −18.71 dB for the S–WLS sparse algorithm. The amplitude error is reduced by 1.0 dB compared to the un–sparse–processed MM algorithm. Therefore, it can be concluded that the proposed MRMM sparse algorithm is highly effective with high error accuracy.

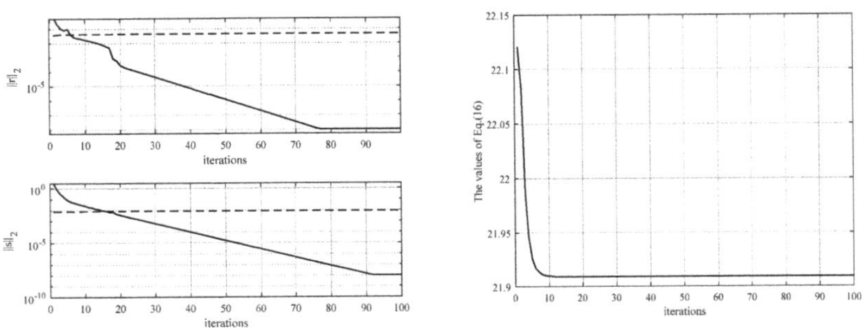

(a) (b)

Figure 3. Convergence of the S−ADMM algorithm: (**a**) shows the convergence of $\|r\|_2$ and $\|s\|_2$; and (**b**) shows the convergence of the cost function given by Equation (16).

Table 1. Comparison of error performance of different algorithms.

	$M = 5$		$M = 7$		$M = 9$		$M = 11$	
	ε_1	ε_2	ε_1	ε_2	ε_1	ε_2	ε_1	ε_2
MM	−19.83	0.0067	−20.01	0.0061	−18.69	0.0085	−23.24	0.0047
S–WLS	−17.74	0.01	−20.78	0.0054	−18.71	0.0072	−20.44	0.0061
MRMM	−18.37	0.0097	−19.45	0.0080	−19.81	0.0066	−21.98	0.0058

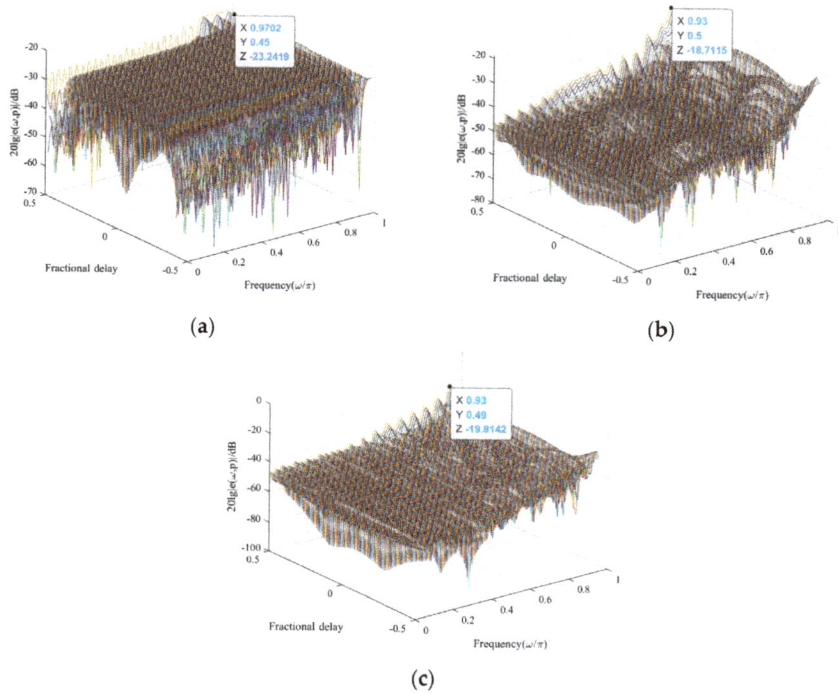

Figure 4. Amplitude–frequency response error plots for different algorithms. (**a**) is the MM algorithm; (**b**) is the S–WLS algorithm; and (**c**) is the MRMM algorithm.

Next, the amplitude–frequency response error and phase–frequency response error of the designed VFD filter were analyzed. Experimental results are given for the number of sub–filters $M = 9$ and the initial order of sub–filters $N = 30$. The magnitude–frequency response plots of different algorithms are given as shown in Figure 5, where it can be seen that the normalized magnitude of the un–sparse standard MM algorithm is close to one between the band range [0,0.9], and the normalized magnitude of the sparse–processed S–WLS algorithm and the MRMM algorithm fluctuates around one between the band range [0,0.9], but the range of fluctuation is within acceptable limits. In the bandwidth range [0.9,1], the maximum amplitude fluctuation of the MRMM algorithm is only 0.05, while the maximum amplitude fluctuation of the S–WLS algorithm in this range is 0.12. Next, Figure 6 shows the phase–frequency curves of the different algorithms where fractional delay $p = 0.3$; the maximum phase fluctuation of the S–WLS algorithm is 0.004; and the maximum phase fluctuation of the MRMM algorithm is only 0.002. Therefore, it can be proved that the MRMM algorithm proposed in this paper has more stable amplitude–frequency characteristics than the existing sparse algorithms.

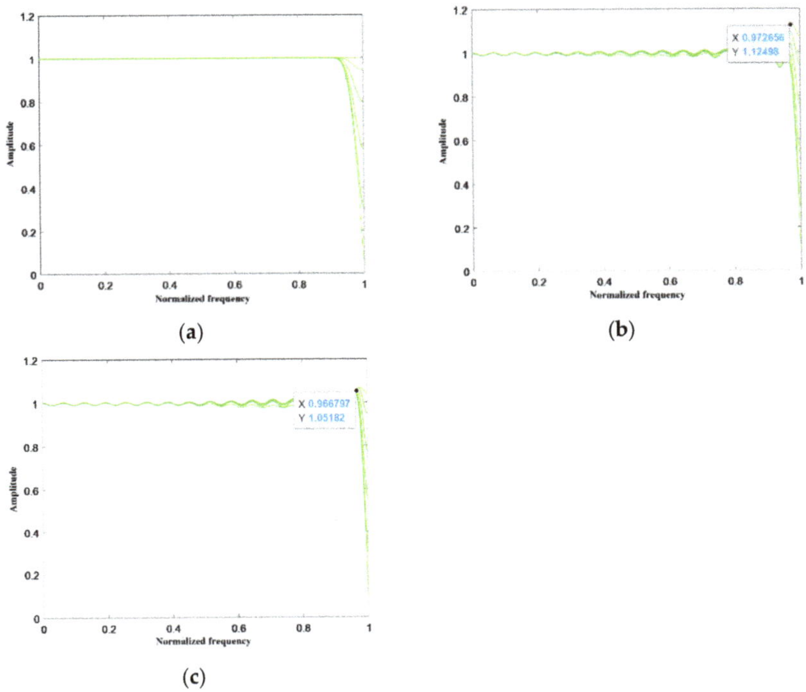

Figure 5. Amplitude–frequency response plots of different algorithms. (**a**) is the MM algorithm; (**b**) is the S–WLS algorithm; and (**c**) is the MRMM algorithm.

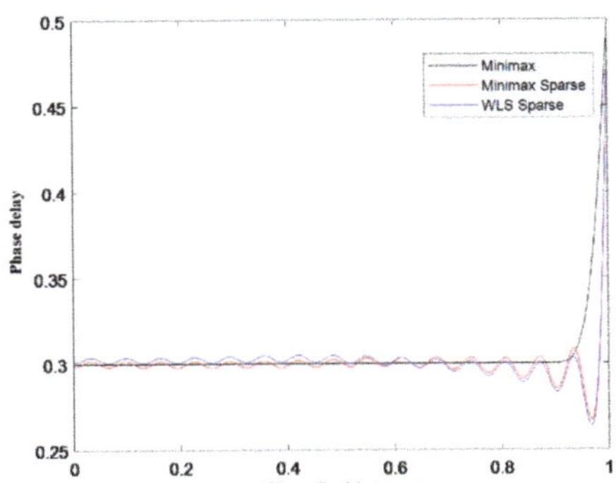

Figure 6. Phase–frequency plot for different algorithms at a fractional delay $p = 0.3$.

To evaluate the sparsity of the MRMM algorithm and the effect of the sparse design on system complexity, the sparsity rate is defined as $\eta = \frac{P_s}{P} \times 100\%$ where P_s represents the number of zero coefficients after being sparse and $P = 2(N+1)(M+1)$ represents the number of filter coefficients under the initial setting. The experimental parameters are M = 9, N = 30, and P = 620. The results of the sparsity rate η and operation time Tc comparison of different algorithms are given in Table 2. The data in the table show that at M = 5 and M = 7, the sparsity rate of the MRMM algorithm is 66.67% and 75.40%, while the sparsity rate of the S–WLS algorithm is 69.89% and 75.80%, respectively. It can be seen that the sparsity of the S–WLS algorithm is a little bit better when the number of sub–filters is small, and the MRMM algorithm is better when the number of sub–filters is large. However, the sparsity of the sparse algorithms MRMM and S–WLS is much greater than that of the standard MM algorithm, regardless of the experimental conditions. In terms of computation time, we can see that the MRMM algorithm exhibits higher computational efficiency, especially when dealing with more sub–filters, and the MRMM algorithm is able to complete the computation faster. Therefore, it can be shown the sparsity of the MRMM algorithm designed in this paper fully satisfies our desired design.

Table 2. Comparison of sparsity and operation time of different algorithms.

	$M = 5$		$M = 7$		$M = 9$		$M = 11$	
	η	*Tc*	η	*Tc*	η	*Tc*	η	*Tc*
MRMM	66.67% (248)	0.16 s	75.40% (374)	0.29 s	82.64% (512)	0.61 s	85.75% (638)	0.73 s
LCMM	37.97% (139)	0.62 s	21.77% (108)	2.05 s	31.93% (198)	2.40 s	32.52% (242)	3.67 s
S–WLS	69.89% (260)	0.14 s	75.80% (376)	0.26 s	81.29% (504)	0.77 s	85.21% (634)	0.92 s

Next, the system complexity of the VFD filter designed by the sparse MRMM algorithm is evaluated. The presence of zero coefficients at lower orders saves the use of corresponding multipliers, and the presence of zero coefficients at higher orders saves the corresponding multipliers and adders. By statistically analyzing the coefficient matrix elements, it can be determined that the filter designed by the standard MM algorithm requires 844 multipliers and 421 adders, which results in extremely high complexity for the system. However, the multipliers and adders required by the sparse MRMM algorithm are 240 and 119, corresponding to a reduction of 71.56% and 71.73%, respectively. It can be concluded that the use of multipliers and adders is significantly reduced by the MRMM sparse design algorithm, which effectively reduces system complexity.

The above experiment analysis proves that the improved VFD filter based on the MRMM model can achieve our purpose well. We then used it to correct the aperture effect in a wideband array. Assume that the angle of incidence of the signal is 10°; the center frequency is 1.2 GHz; and the $f_H : f_L = 3 : 1$. The uniformly distributed array element spacing is $0.5\lambda_{\min}$, and the number of full arrays is 30. Then, the number of arrays after sparse reconstruction is 20, as shown in Figure 7, which is designed according to [24].

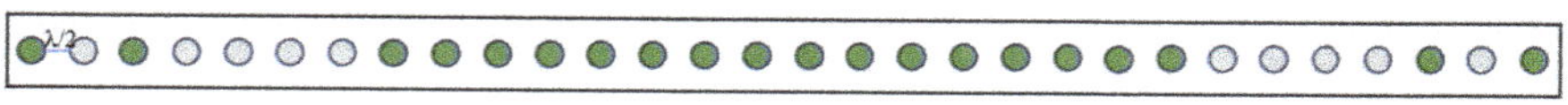

Figure 7. Sparse array location distribution.

Compared with the conventional wideband aperture effect correction for a TDL structure [24], our proposed MRMM–VFD–based wideband aperture effect correction method can keep the filter coefficients unchanged under different scan angles, and only the delay parameter needs to be adjusted. Compared with the LCMM–VFD filter, our proposed MRMM–VFD can significantly reduce the use of multipliers and adders and decrease system complexity. Figure 8a shows the aperture crossing phenomenon of a wideband sparse array, and it can be seen that serious pointing deviations occur in the received signals of different arrays. Figure 8b shows that the received signals of each array are aggregated to the same distance unit after aperture correction using the improved VFD filter, which completely solves beam dispersion. The antenna orientation diagrams of the wideband sparse array with different frequency signals are given in Figure 9. It can be seen above that it is effective to use the proposed VFD filter for wideband sparse arrays during aperture correction.

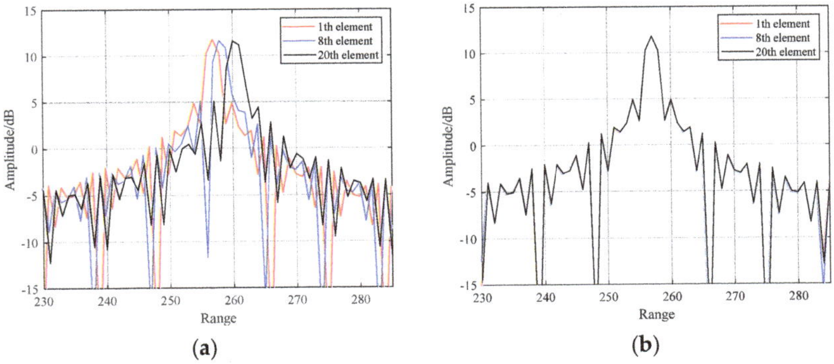

Figure 8. Wideband sparse array antenna orientation diagram. (**a**) is the result of aperture crossing; and (**b**) is the result of aperture correction.

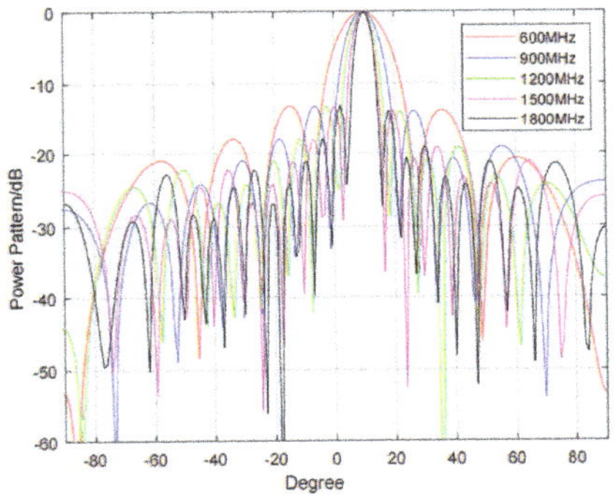

Figure 9. Wideband sparse array antenna orientation diagram.

6. Conclusions

In order to reduce the system complexity of the VFD FIR filter, a new MRMM sparse algorithm under a great–minimal criterion is proposed. The error function is constrained by introducing $L2$ and $L1$ regularizations. To solve this nonconvex model, an improved S–ADMM algorithm is proposed, which uses sequential alternation to iteratively update multiple soft–threshold problems. Experimental results show that the proposed MRMM algorithm completely satisfies the desired error value and sparsity. The use of multipliers and adders is reduced by 71.56% and 71.73%, respectively, leading to a significant decrease in system complexity. Simulation results confirm that the MRMM–VFD effectively corrects AFT in a wideband sparse array.

Author Contributions: Conceptualization, J.G. and M.X.; methodology, J.G., M.X. and M.S.; software, M.S. and W.Z.; validation, J.G., M.X., M.S. and W.Z.; formal analysis, J.G. and M.X.; investigation, J.G. and M.X.; resources, J.G., M.X. and M.S.; data curation, J.G. and M.X.; writing—original draft preparation, J.G.; writing—review and editing, M.X., M.S. and W.Z.; visualization, J.G.; supervision, J.G. and M.X.; project administration, J.G. and M.X.; funding acquisition, J.G., M.X., M.S. and W.Z. All authors have read and agreed to the published version of the manuscript.

Funding: This research was funded by the Natural Science Foundation of China [No.62271190], and Science and Technology on Electronic Information Control Laboratory [No.20220003].

Data Availability Statement: The data used to support the findings of this study are available from the corresponding author upon request.

Acknowledgments: We sincerely thank all the reviewers for their patient reviews and valuable suggestions on this article.

Conflicts of Interest: The authors declare that the research was conducted in the absence of any commercial or financial relationships that could be construed as a potential conflict of interest.

References

1. Edussooriya, C.U.S.; Bruton, L.T.; Naeini, M.A.; Agathoklis, P. Using 1-D Variable Fractional-Delay Filters to Reduce the Computational Complexity of 3-D Broadband Multibeam Beamformers. *IEEE Trans. Circuits Syst. II Express Briefs* **2014**, *61*, 278–283. [CrossRef]
2. Canese, L.; Cardarilli, G.C.; Di Nunzio, L.; Fazzolari, R.; Giardino, D.; Re, M.; Spanó, S. Efficient Digital Implementation of a Multirate-Based Variable Fractional Delay Filter for Wideband Beamforming. *IEEE Trans. Circuits Syst. II Express Briefs* **2023**, *70*, 2231–2235. [CrossRef]
3. Kokila, R.; Chithra, K.; Dhilsha, R. Wideband Beamforming Using Modified Farrow Structure FIR Filtering Method for Sonar Applications. In Proceedings of the 2019 International Symposium on Ocean Technology (SYMPOL), Ernakulam, India, 11–13 December 2019.
4. Wang, J.; Cai, D.; Lei, P. Aperture effect influence and analysis of stepped frequency wideband phased array radar. In Proceedings of the IEEE CIE International Conference on Radar, Chengdu, China, 24–27 October 2011.
5. Selva, J. An Efficient Structure for the Design of Variable Fractional Delay Filters Based on the Windowing Method. *IEEE Trans. Signal Process.* **2008**, *56*, 3370–3775. [CrossRef]
6. Deng, T.-B. Decoupling Minimax Design of Low-Complexity Variable Fractional-Delay FIR Digital Filters. *IEEE Trans. Circuits Syst. I Regul. Pap.* **2011**, *58*, 2398–2408. [CrossRef]
7. Deng, T.-B. Minimax Design of Low-Complexity Even-Order Variable Fractional-Delay Filters Using Second-Order Cone Programming. *IEEE Trans. Circuits Syst. II Express Briefs* **2011**, *58*, 692–696. [CrossRef]
8. Huang, Y.-D.; Pei, S.-C.; Shyu, J. WLS Design of Variable Fractional-Delay FIR Filters Using Coefficient Relationship. *IEEE Trans. Circuits Syst. II Express Briefs* **2009**, *56*, 220–224. [CrossRef]
9. Deng, T.-B.; Qin, W.-B.; Qin, W. Coefficient relation-based minimax design and low-complexity structure of variable fractional-delay digital filters. *Signal Process.* **2012**, *93*, 923–932. [CrossRef]
10. Farrow, C.W. A continuously variable digital delay element. In Proceedings of the IEEE International Symposium on Circuits and Systems, Espoo, Finland, 7–9 June 1988.
11. Deng, T.-B. Symmetric Structures for Odd-Order Maximally Flat and Weighted-Least-Squares Variable Fractional-Delay Filters. *IEEE Trans. Circuits Syst. I Regul. Pap.* **2007**, *54*, 2718–2732. [CrossRef]
12. Jiang, A.M.; Kwan, H.K.; Zhu, Y.P.; Liu, X.F.; Xu, N.; Tang, Y.B. Design of sparse FIR filters with joint optimization of sparsity and filter order. *IEEE Trans. Circuits Syst. I Regul. Pap.* **2015**, *62*, 195–204. [CrossRef]
13. Chen, W.Q.; Huang, M.; Lou, X. Design of Sparse FIR Filters With Reduced Effective Length. *IEEE Trans. Circuits Syst. I Regul. Pap.* **2019**, *66*, 1496–1506. [CrossRef]
14. Lu, W.-S.; Hinamoto, T. Variable fractional delay FIR filters with sparse coefficients. In Proceedings of the 2012 IEEE International Symposium on Circuits and Systems (ISCAS), Seoul, Republic of Korea, 20–23 May 2012.
15. Zhou, W.J.; Shen, M.W.; Xu, M.; Han, G.; Zhang, Y. Sparsity-optimised farrow structure variable fractional delay filter for wideband array. *IET Signal Process.* **2023**, *17*, e12228. [CrossRef]
16. Shentu, X.; Lai, X.; Wang, T.; Cao, J. Efficient ADMM-Based Algorithm for Regularized Minimax Approximation. *IEEE Signal Process. Lett.* **2023**, *30*, 210–214. [CrossRef]
17. Chiu, W.-Y.; Chen, B.-S.; Yang, C.-Y. Robust Relative Location Estimation in Wireless Sensor Networks with Inexact Position Problems. *IEEE Trans. Mob. Comput.* **2012**, *11*, 935–946. [CrossRef]
18. Nakamoto, M.; Aikawa, N. Minimax Design of Sparse IIR Filters Using Sparse Linear Programming. *IEICE Trans. Fundam. Electron. Commun. Comput. Sci.* **2012**, *104*, 1006–1018. [CrossRef]
19. Nordebo, S.; Claesson, I. Minimum norm design of two-dimensional weighted Chebyshev FIR filters. *IEEE Trans. Circuits Syst. II Analog Digit. Signal Process.* **1997**, *44*, 251–253. [CrossRef]
20. Lu, W.-S.; Hinamoto, T. Digital filters with sparse coefficients. In Proceedings of the 2010 IEEE International Symposium on Circuits and Systems, Paris, France, 30 May–2 June 2010.
21. Baran, T.; Wei, D.; Oppenheim, A.V. Linear Programming Algorithms for Sparse Filter Design. *IEEE Trans. Signal Process.* **2010**, *58*, 1605–1617. [CrossRef]
22. Matsuoka, R.; Kyochi, S.; Ono, S.; Okuda, M. Joint Sparsity and Order Optimization Based on ADMM With Non-Uniform Group Hard Thresholding. *IEEE Trans. Circuits Syst. I Regul. Pap.* **2018**, *65*, 1602–1613. [CrossRef]
23. Yanamshetti, R.; Bharatkar, S.S.; Chatterjee, D.; Ganguli, A.K. A hybrid fuzzy based loss minimization technique for fast efficiency optimization for variable speed induction machine. In Proceedings of the 2009 6th International Conference on Electrical Engineering/Electronics, Computer, Telecommunications and Information Technology, Chonburi, Thailand, 6–9 May 2009.

24. Hawes, M.B.; Liu, W. Sparse array design for wideband beamforming with reduced complexity in tapped delay-lines. *IEEE/ACM Trans. Audio Speech Lang. Process.* **2014**, *22*, 1236–1247. [CrossRef]
25. Zhao, R.; Hong, X. Matrix-Based Algorithms for the Optimal Design of Variable Fractional Delay FIR Filters. *IEEE Trans. Circuits Syst. I Regul. Pap.* **2019**, *66*, 4740–4752. [CrossRef]

sensors

MDPI

Communication

A Novel Generalized Nested Array MIMO Radar for DOA Estimation with Increased Degrees of Freedom and Low Mutual Coupling

Zhongtian Yang [1], Zhengyang Bi [2], Ye Chen [3,*] and Honghao Hao [3]

[1] School of Energy and Power Engineering, Beihang University, Beijing 102200, China; buaayztian@buaa.edu.cn
[2] College of Advanced Technology, Xi'an Jiaotong-Liverpool University, Suzhou 215000, China; zhengyang.bi22@student.xjtlu.edu.cn
[3] College of Electronic Information Engineering, Nanjing University of Aeronautics and Astronautics, Nanjing 210016, China; haowenjie1025@163.com
* Correspondence: amberchenye@nuaa.edu.cn

Abstract: In array signal processing, the mutual coupling among physical sensors can inevitably affect the estimation of the direction of arrival (DOA). Despite the fact that multiple-input and multiple-output (MIMO) radar can provide greater degrees of freedom (DOFs), the influence of mutual coupling is largely overlooked in many current MIMO radar designs. To tackle this issue, we propose the utilization of a generalized nested array (GNA) in transmitter array and we introduce an expansion factor into the nested array in the receiver array. Thereby, a novel GNA-MIMO radar is put forward. The proposed MIMO radar offers $O(N^4)$ consecutive DOFs with N sensors and avoids the adverse effects of high mutual coupling caused by closely located sensors. Furthermore, we derive the closed-form expressions for the position of physical sensors and the attainable consecutive DOFs of the proposed MIMO radar. Through simulation experiments, we demonstrate the superior accuracy of the proposed MIMO configuration in DOA estimation and angle resolution under the condition of mutual coupling effect.

Keywords: multiple-input multiple-output (MIMO) radar; nested array; direction of arrival estimation; degree of freedom; mutual coupling

1. Introduction

In the field of array signal processing, direction of arrival (DOA) estimation is a crucial topic with significant importance in wireless communication, radar and medical imaging, etc., [1–4]. Multi-input multi-output (MIMO) radar is an innovative radar system that was initially proposed in 2003 [5]. In the past decades, MIMO radar has drawn substantial concern and is becoming an increasingly hot topic [3,6]. One of the advantages of MIMO radar is the ability to create a virtual array with large aperture and increased degrees of freedom (DOFs) by utilizing multiple transmitter and receiver antennas [7]. The number of DOFs is one of the key parameters to measure the performance of DOA in MIMO radar systems, which represents the ability of radar to distinguish and process multiple targets simultaneously. DOFs are mainly determined by the number of antennas, the distance between antennas and the arrangement of sensors. By increasing DOFs, the resolution, detection accuracy and anti-interference ability of the radar system can be significantly improved. However, the current research on MIMO radar systems mainly focuses on ways to achieve higher DOFs while few of them study the mutual coupling effect between physical sensors. The impact of mutual coupling cannot be neglected in practice and can degrade the performance of DOA estimation, especially when physical sensors are closely located [8].

Various design methods have been put forward to design MIMO radar array configurations with better properties. Initially, the design of minimal redundancy (MR) MIMO radar was proposed to reduce sensor redundancy [9–12]. Nonetheless, this method suffers from high computational complexity, leading to increased costs. Subsequently, the advancement of nested array (NA) [13], coprime array (CA) [14] and other sparse linear arrays [15–17] have offered a new perspective in MIMO radar system design. For example, Qin et al. proposed a design method for nested MIMO radar [18] and co-prime MIMO radar [19] through the utilization of two subarrays of NA (or CA) in the transmitter and receiver arrays, respectively. However, the above-mentioned MIMO radar has the limitation of restricted DOFs, the maximal DOFs of which can only reach $O(N^2)$ with N physical sensors. To increase DOFs, the concept of difference co-array of sum co-array (DCSC) [12] as well as the generalized sum co-array of difference co-array (GSDC) [13] were proposed. These two methods gained significant attention for the ability to achieve $O(N^4)$ DOFs with only N physical sensors. Zheng et al. proposed a nested MIMO radar with a sparsely extended array geometry, the transmitter array of which is composed of extended nested arrays by introducing an expansion factor [20]. Likewise, a generalized co-prime MIMO radar was proposed with generalized co-prime arrays by introducing an expansion factor to the transmitter array [21]. Later in [22], a flexible MIMO radar was proposed to enhance the element spacing of transmitter and receiver arrays by generalizing the coprime expansion factors. It is worth mentioning that the above conventional nested and co-prime MIMO radars represent special cases of flexible MIMO radar. Additionally, Zhang et al. proposed a novel generalized nested MIMO radar by replacing the nested array with an extended nested array to achieve more DOFs [23]. Furthermore, to gain more attainable DOFs, Shaikh et al. devised a generalized multi-level MIMO radar configuration [24].

However, the above-mentioned MIMO radar systems neglected the influence of mutual coupling. Consequently, despite the great number of DOFs, these systems suffer from significant degradation in DOA performance caused by severe mutual coupling in practice. In [25,26], a UCA-MIMO radar was proposed by using unfolded co-prime arrays (UCA) in the transmitter and receiver arrays. Although this design only achieves $O(N^2)$ DOFs, the advantage lies in the sparsely located sensors. In [27], by using uniform linear arrays in the transmit-receiver array and introducing coprime factors, a generalized two-level nested array (GTNA) MIMO radar was proposed. Benefiting from the coprime factor, the element spacing continuously increases while the consecutive DOFs gradually decrease, thus reaching a balance between a great number of consecutive DOFs and a weak mutual coupling effect. However, the optimal selection of coprime factors was not discussed. In [28], a UGCA-MIMO radar was proposed by utilizing an unfolded generalized co-prime array (UGCA) in the transmitter and receiver arrays. However, no closed-form expression was provided. To address these issues and limitations, this paper puts forward the design of a GNA-MIMO radar. The proposed structure utilizes a GNA in the transmitter array and a prototype NA with a sparse expansion factor in the receiver array. The main contributions of this paper are as follows:

(1) The proposed GNA-MIMO radar reaches the balance between both a great number of consecutive DOFs and a weak mutual coupling leakage effect. Specifically, it achieves $O(N^4)$ consecutive DOFs with only N physical sensors with no need for closely located physical sensors in the transmitter or receiver arrays. This achievement is noteworthy as it surpasses the capabilities of most previously devised MIMO radars.

(2) We derive the closed-form expressions for the location of physical sensors and the attainable consecutive DOFs for the proposed GNA-MIMO radar. Meanwhile, we present the design principle for GNA-MIMO radar in Section 3.3. In contrast, the UGCA-MIMO radar designed in [29] lacks closed-form expressions and is not suitable for all possible numbers of sensors. Simulation results also prove that the proposed GNA-MIMO radar has a weaker coupling effect.

The paper is outlined as follows: Section 2 introduces the data model of the monostatic MIMO radar. The GNA-MIMO radar geometry is proposed in Section 3. In Section 4, sufficient theoretical analysis is provided. Section 5 provides lots of numerical results and Section 6 concludes.

Throughout this paper, the upper-case bold and lower-case bold characters, respectively, denote matrices and vectors; $(\cdot)^T$, $(\cdot)^*$ and $(\cdot)^H$ denote transpose, conjugate and conjugate transpose; $\odot$ and $\otimes$ represent Khatri–Rao product and Kronecker product; $\mathbf{I}_N$ denotes an $N \times N$ identity matrix; $E[\cdot]$ denotes the expectation operator; $diag(\cdot)$ denotes the diagonal operation; $vec(\cdot)$ stands for the vectorization operator.

2. Signal Model

A narrow-banded monostatic MIMO radar system is composed of an M-element transmitter array and an N-element receiver array. The transmitter array emits multiple orthogonal signals while the receiver array captures the reflected signals.

Assuming that there are K far-field incoherent targets, where θ_k denotes the direction of the k-th target. Then the MIMO steering vector $\mathbf{a}(\theta_k)$ equals to [20]

$$\mathbf{a}(\theta_k) = \mathbf{a}_r(\theta_k) \otimes \mathbf{a}_t(\theta_k), \tag{1}$$

where $\mathbf{a}_r(\theta_k)$ and $\mathbf{a}_t(\theta_k)$ are the steering vectors of the receiver and transmitter corresponding to the k-th target, respectively. And $\mathbf{a}_r(\theta_k)$ and $\mathbf{a}_t(\theta_k)$ can be given as [20]

$$\mathbf{a}_r(\theta_k) = [e^{\frac{j2\pi d_1 sin\theta_k}{\lambda}}, \cdots, e^{\frac{j2\pi d_N sin\theta_k}{\lambda}}]^T, \tag{2}$$

$$\mathbf{a}_t(\theta_k) = [e^{\frac{j2\pi d_1' sin\theta_k}{\lambda}}, \cdots, e^{\frac{j2\pi d_M' sin\theta_k}{\lambda}}]^T, \tag{3}$$

where λ denotes the wavelength, $d_n(n = 1, 2, \cdots, N)$ and $d'_m(m = 1, 2, \cdots, M)$ stands for the sensor locations in the receiver and transmitter arrays.

Further, $\mathbf{a}(\theta_k)$ can be written as [21]

$$\begin{aligned}\mathbf{a}(\theta_k) = [&e^{j2\pi(d_1+d_1')sin\theta_k/\lambda}, e^{j2\pi(d_1+d_2')sin\theta_k/\lambda} \\ &, \cdots, e^{j2\pi(d_2+d_1')sin\theta_k/\lambda}, e^{j2\pi(d_2+d_2')sin\theta_k/\lambda} \\ &, \cdots, e^{j2\pi(d_N+d_M')sin\theta_k/\lambda}],\end{aligned} \tag{4}$$

where the MIMO steering vector $\mathbf{a}(\theta_k)$ is a $MN \times 1$ steering vector.

Subsequently, taking mutual coupling into account, the signal model of the received signal can be formulated as [28]

$$\begin{aligned}\mathbf{X}(t) &= \mathbf{C}_t \otimes \mathbf{C}_r \sum_{k=1}^{K} \mathbf{a}_t(\theta_k) \otimes \mathbf{a}_r(\theta_k)\mathbf{s}(t) + \mathbf{n}(\mathbf{t}) \\ &= \mathbf{CAs}(t) + \mathbf{n}(t),\end{aligned} \tag{5}$$

where $\mathbf{s}(t) = [s_1(t), s_2(t), \ldots, s_K(t)]$ represents the signal source vector, $\mathbf{n}(t)$ represents the zero-mean Gaussian white noise vector, $\mathbf{A} = [\mathbf{a}(\theta_1), \cdots, \mathbf{a}(\theta_K)]$ represents the steering matrix and $\mathbf{a}(\theta_k) = \mathbf{a}_t(\theta_k) \otimes \mathbf{a}_r(\theta_k)$ is the steering vector of the k-th signal. The mutual coupling matrix $\mathbf{C} = \mathbf{C}_t \otimes \mathbf{C}_r$. The mutual coupling matrice $\mathbf{C}$ can be approximated as [15]

$$\mathbf{C}_{ij} = \begin{cases} c_{|p_i - p_j|}, & |p_i d - p_j d| \le B \\ 0, & |p_i d - p_j d| > B \end{cases} \tag{6}$$

where $\mathbf{C}_{ij}$ denotes the element located in the i-th row and j-th column of the matrix $\mathbf{C}$ and $p_i d$ denotes the position of i-th sensor (i = 1, 2, … N). $1 = c_0 > |c_1| > \cdots > |c_B| > |c_{B+1}| = 0$, $c_l = c_1 e^{-j(l-1)\pi/8}/l$ for $2 \le B \le 100$ [30]. Subsequently, the coupling leakage can be measured by the parameter [30]

$$\mathbf{L}(M) = \frac{||\mathbf{C} - \mathbf{I}||_F}{||\mathbf{C}||_F} \tag{7}$$

where $\mathbf{I}$ represents the identity matrix with the same order as the matrix $\mathbf{C}$.

The covariance matrix of the received signal can be calculated as [20]

$$\mathbf{R}_x = E[\mathbf{X}(t)\mathbf{X}^H(t)] = \mathbf{A}\mathbf{R}_s\mathbf{A}^H + \sigma_n^2\mathbf{I}_{N^2}, \tag{8}$$

where $\mathbf{R}_s = E[\mathbf{s}(t)\mathbf{s}^H(t)] = diag([\sigma_1^2, \ldots, \sigma_K^2])$ and σ_k^2 denotes the power of the k-th source. In practice, the covariance matrix is calculated through a certain number of snapshots, $\hat{\mathbf{R}}_x = (1/L)\sum_{t=1}^{L}\mathbf{X}(t)\mathbf{X}^H(t)$, where L is the number of snapshots.

Vectorizing the covariance matrix in (8) yields [20]

$$\mathbf{z} = vec(\mathbf{R}_x) = vec(\mathbf{A}_0\mathbf{p}) + \sigma_n^2 vec(\mathbf{I}_{N^2}), \tag{9}$$

where $\mathbf{A}_0 = \mathbf{A}^* \odot \mathbf{A} = [\mathbf{a}^*(\theta_1) \otimes \mathbf{a}(\theta_1), \mathbf{a}^*(\theta_2) \otimes \mathbf{a}(\theta_2), \cdots, \mathbf{a}^*(\theta_K) \otimes \mathbf{a}(\theta_K)]$ is a new steering matrix including a lot of virtual array elements; $\mathbf{p} = [\sigma_1^2, \ldots, \sigma_K^2]^T$. From (9), the vector $\mathbf{z}$ indicates a new received signal corresponding with the new steering matrix $\mathbf{A}_0$ with a single snapshot.

3. Generalized Nested Array MIMO Radar Configuration

In this section, we present a concise review of sum co-array and difference co-array. Second, the reformulation of GNA is shown. Later, the design process of the Generalized Nested Array MIMO radar is presented. At last, the closed-form expression of consecutive virtual array elements and the available consecutive DOFs are derived in detail.

3.1. Review of Sum Co-Array and Difference Co-Array

Definition 1. *The position set of sensors for a given $M-$ element array is denoted by set $\mathbb{S}$:*

$$\mathbb{S} = \{p_1, p_2, \ldots, p_M\}d.$$

where M represents the number of elements and d denotes the unit element spacing, $d = \lambda/2$. In subsequent representations in this paper, we omit the parameter d.

Definition 2. *The sum co-array generated by the arrays $\mathbb{S}_a$ and $\mathbb{S}_b$ can be defined by the unique elements in the set $\mathbb{S}_{sum}$, which can be written as [18]*

$$\mathbb{S}_{sum} = \{u + u' | u \in \mathbb{S}_a, u' \in \mathbb{S}_b\}.$$

Definition 3. *The difference co-array generated by the array $\mathbb{S}_u$ is symmetric and can be similarly defined by the unique elements in the set $\mathbb{S}_{diff}$, which can be given as [31]*

$$\mathbb{S}_{diff} = \{u - u' | u, u' \in \mathbb{S}_u\}.$$

Definition 4. *Given an array $\mathbb{S}$, the number of elements in the consecutive segment of its difference co-array $\mathbb{S}_{diff}$ is defined as "consecutive DOFs".*

For instance, the ULAs $\mathbb{S}_a = \{0, 1, 2\}$ and $\mathbb{S}_b = \{0, 3\}$ can obtain the sum co-array $\mathbb{S}_{sum} = \{0, 1, 2, 3, 4, 5\}$ in the end. $\mathbb{S}_u = \{1, 2, 3, 6\}$ can finally obtain a uniform liner difference co-array $\mathbb{S}_{diff} = \{-5, -4, -3, -2, -1, 0, 1, 2, 3, 4, 5\}$. In this case, the consecutive DOFs of S_{diff} is 11 (From -5 to 5).

3.2. Reformulation of Generalized Nested Array

A generalized nested array (GNA) consists of two sparse uniform linear subarrays [15]. One subarray consists of M_1 elements with element-spacing αd_0 while the other subarray contains M_2 elements with element-spacing βd_0, where d_0 is the unit sensor spacing and will be omitted in the subsequent text. Here, α and β are flexible co-prime factors satisfying $1 \leq \alpha \leq M_2, 1 \leq \beta \leq M_1 + 1$. The position set of GNA elements can be represented as

$$\mathbb{S}_{GNA} = \{1, 1+\alpha, \cdots, 1+\alpha M_1, 1+\alpha M_1+\beta, \\ \cdots, 1+\alpha M_1+\beta(M_2-1)\}. \tag{10}$$

In [15], it has been demonstrated that the DOFs of GNA are related to the variables α, β, M_1 and M_2. Notably, the maximum DOFs of GNA are attained when $M_1 = M_2$ or $M_2 = M_1 + 1$ and one of the two factors α or β reaches its maximal value. Under this circumstance, the difference co-array of GNA is non-consecutive and results in a decrease in the mutual coupling leakage rate.

3.3. Design Principle for GNA-MIMO Radar

According to the GSDC concept [21], the virtual array elements of a MIMO radar can be computed by the operation sum co-array of the transmitter's difference co-array and the receiver's difference co-array. This approach allows us to utilize a GNA in the transmitter array with the maximum possible element spacing and an NA with an extension factor γ in the receiver array. Despite the non-continuous nature of their individual difference co-arrays, the sum co-array of these two difference co-arrays remains continuous. Consequently, this design strategy offers not only a large number of consecutive DOFs but also mitigates the mutual coupling effect.

The configuration of the proposed GNA-MIMO radar involves two steps:

(1) For the transmitter array, we employ an M-element GNA. The optimal values for variables are provided in Table 1 [15]. When M is even, we take the maximum value of α, β, that is $\alpha = M_2 = M/2$, $\beta = M_1 + 1 = M/2 + 1$. In this case, $\beta = \alpha + 1$, α and β satisfy the co-prime relation. However, if we take the maximum value of α, β when M is odd, α and β are not co-prime. So, we think $\alpha = M_2 - 1 = (M-1)/2$. In this case, $\beta = \alpha + 1$, α and β also satisfy the co-prime relation. The difference co-array for the transmitter array, denoted as $\mathbb{D}_T$, can be written as

$$\begin{aligned} \mathbb{D}_T &= \mathbb{D}_T^+ \cup \mathbb{D}_T^- \\ \mathbb{D}_T^+ &= \{n - n' | n \geq n', n, n' \in \mathbb{S}_T\} \\ &= \{\alpha m_1 + \beta m_2 | 0 \leq m_1 \leq M_1, 0 \leq m_2 \leq M_2 - 1\}. \end{aligned} \tag{11}$$

where $\mathbb{D}_T^+$ and $\mathbb{D}_T^-$ denote the non-negative and negative values in $\mathbb{D}_T$, respectively.

Table 1. The Optimal Variables Values of M_1,M_2 and α,β.

When M Is	The Optimal M_1, M_2	The Optimal α,β
Even	$M_1 = M_2 = M/2$	$\alpha = M/2, \beta = M/2+1$
Odd	$M_1 = (M-1)/2, M_2 = (M+1)/2$	$\alpha = (M-1)/2, \beta = (M+1)/2$

(2) For the receiver array, we utilize an N-element NA with an extension factor $\gamma = M_2(M_1+1)$. If N is even, the number of elements for the two subarrays is $N_1 = N_2 = N/2$, respectively, otherwise, if N is odd, $N_1 = (N-1)/2$ and $N_2 = (N+1)/2$. The difference co-array for the receiver array, denoted as $\mathbb{D}_R$, can be presented as [13]

$$\mathbb{D}_R = \gamma\{-(N_2(N_1+1)-1), \cdots, 0, 1, \cdots, N_2(N_1+1)-1\}. \tag{12}$$

In Figure 1, we present an instance of the proposed GNA-MIMO radar with a six-element transmitter array and a four-element receiver array. The position of the array element is determined based on the aforementioned design rules. Considering the symmetric property of difference co-array, only the non-negative part of the GSDC is plotted. Figure 1 illustrates that although elements in the transmitter and receiver arrays are sparsely located, the GSDC still has a long consecutive segment.

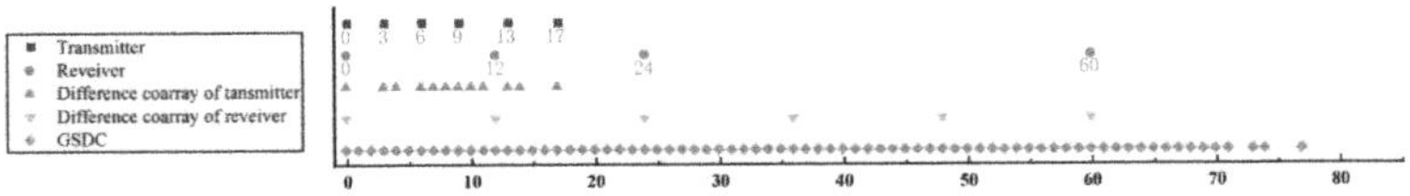

Figure 1. An example of a proposed radar array with a six-element transmitter array and a four-element receiver array. Only the non-negative part of the GSDC is given. The squares in the Transmitter and circles in Receiver are the actual positions of physical sensors while the others are virtual elements.

3.4. Properties of the GNA-MIMO Radar

Proposition 1. *The closed-form expression of attainable consecutive DOFs for the proposed GNA-MIMO radar is*

$$f = 2(M_2(M_1+1))(N_2(N_1+1)) - 1. \tag{13}$$

Proof. The $\mathbb{D}_T^+$ in (11) has the following properties:

(a) There are $M_2(M_1+1)$ distinct integers in the set $\mathbb{D}_T^+$.

(b) $0 \leq \alpha m_1 + \beta m_2 \leq \alpha M_1 + \beta(M_2 - 1)$.

(c) There are not such two values in $\mathbb{D}_T^+$ that differ by a distance of $M_2(M_1+1)$. In other words, the inequality $\alpha m_1 + \beta m_2 - (\alpha m_1' + \beta m_2') \neq M_2(M_1+1)$ holds true.

Dividing both sides of the inequation by β, the expression $\frac{\alpha}{\beta}(m_1 - m_1') \neq M_2 + (m_2 - m_2')$. We can see that $\frac{\alpha}{\beta}(m_1 - m_1')$ must be a decimal number except for $m_1 = m_1'$ while the right side of the inequality is a positive integer. Therefore, the inequality definitely holds true.

(d) The set $\mathbb{S}' = \{\mathrm{mod}(\alpha m_1 + \beta m_2, M_2(M_1+1)) | 0 \leq m_1 \leq M_1, 0 \leq m_2 \leq M_2 - 1\}$ contains all integers from 0 to $M_2(M_1+1) - 1$, where mod represents the modulo operation.

From (a) (b) (c), we can conclude that the set $\mathbb{S}'$ also contains $M_2(M_1+1)$ distinct integers in the range $[0, M_2(M_1+1) - 1]$. Therefore, the set $\mathbb{S}'$ contains all integers from 0 to $M_2(M_1+1) - 1$.

(e) The set $\{\mathbb{D}_T^+ - M_2(M_1+1)\} \cup \mathbb{D}_T^+$ contains the set $\mathbb{S}'$ and is consecutive in the range $[0, M_2(M_1+1) - 1]$.

The GSDC for GNA-MIMO $\mathbb{S}_{GSDC}$ has the following properties:

(f) $\mathbb{S}_{GSDC} = \{\mathbb{D}_T^+ \cup \mathbb{D}_T^-\} + \mathbb{D}_R = \{\mathbb{D}_T^- + \mathbb{D}_R\} \cup \{\mathbb{D}_T^+ + \mathbb{D}_R\}$.

(g) The set $\mathbb{D}_R + \mathbb{D}_T^+$ is consecutive in the range $[0, N_2(N_1+1)\gamma - 1]$.

The set $\mathbb{D}_R + \mathbb{D}_T^+$

$$\begin{aligned} &\subseteq \gamma\{-1, 0, \cdots, N_2(N_1+1) - 1\} + \mathbb{D}_T^+ \\ &= \{-\gamma + \mathbb{D}_T^+\} \cup \mathbb{D}_T^+ \cup \cdots \cup \{\gamma(N_2(N_1+1) - 1) + \mathbb{D}_T^+\}. \end{aligned}$$

According to (e), the set $\{-\gamma + \mathbb{D}_T^+\} \cup \mathbb{D}_T^+$ contains $[0, \gamma - 1]$ and the set $\mathbb{D}_T^+ \cup \{\gamma + \mathbb{D}_T^+\}$ contains $[\gamma, 2\gamma - 1]$. By inference, the above set certainly contains

$$\cup[\gamma, 2\gamma - 1] \cup \cdots \cup [(N_2(N_1+1) - 1)\gamma, (N_2(N_1+1))\gamma - 1].$$

After merging, we can conclude that the set $\mathbb{D}_R + \mathbb{D}_T^+$ is consecutive in the range $[0, (N_2(N_1+1))\gamma - 1]$. Based on the property (f), the proposed GNA-MIMO radar ultimately obtains a virtual array from -u to u, where $u = (N_2(N_1+1))\gamma - 1$. Consequently, the consecutive DOFs can be calculated as

$$f = 2u + 1 = 2M_2N_2(M_1+1)(N_1+1) - 1 \tag{14}$$

□

3.5. Optimal M and N

Although the number of physical sensors in the transmitter and receiver arrays can be selected according to the specific requirements, the maximum number of consecutive DOFs can be realized by choosing optimal values of M and N given the number of total physical sensors T. Based on the arithmetic mean–geometric mean inequalities, the optimal values of M and N can be achieved by

$$\begin{cases} M = N = T/2, & \text{if } T \text{ is even} \\ M = (T-1)/2, N = (T+1)/2, & \text{if } T \text{ is odd.} \end{cases} \tag{15}$$

4. Performance Analysis

In this part, we provide sufficient theoretical analysis to demonstrate the advantages of the proposed GNA-MIMO radar. First, four other MIMO radars including UCA-MIMO radar in [25], UGCA-MIMO radar in [28], NA-MIMO radar in [20] and GTNA-MIMO radar in [27] are adopted for comparison. Then, three aspects including the consecutive DOFs, the mutual coupling leakage and the Cramer–Rao Bound (CRB) are considered for analysis.

4.1. The Consecutive DOFs

Usually, the optimal array configuration and design methodology of the MIMO radar should be determined before comparing the consecutive DOFs of the different MIMO radars. Here, for the GTNA-MIMO radar, $\alpha = M_1M_2 - M_1 - M_2$ and $\beta = \alpha + 1$ and the optimal array configurations for the other types of MIMO radar can be found in the corresponding original text.

Table 2 lists the closed-form expression of consecutive DOFs for five types of MIMO radar. In the table, M' and N' represent two coprime numbers in the co-prime MIMO radar, where M_1 and M_2 represent the number of array elements in the transmit array of nested array MIMO radar and N_1 and N_2 represent the number of array sensors in the receive array of nested array MIMO radar. In general, MIMO radar with more DOFs will obtain a more accurate DOA result. Also, from the table, we can observe that the proposed GNA-MIMO radar has relatively high consecutive DOFs and achieves the same consecutive DOFs as the NA-MIMO radar in [20].

Figure 2 depicts the variation of attainable consecutive DOFs with a given number of physical sensors for these five different MIMO radars. It can be observed from Figure 2 that as the number of array elements increases, the consecutive DOFs continuously increase. Given the same number of physical sensors, the proposed GNA-MIMO radar can achieve the maximum consecutive DOFs as the NA-MIMO radar in [20] and exceeds other types of MIMO radar. Additionally, the solid lines in the figure represent that the nested array MIMO radars can take the number of array elements for all integers on the abscissa and the curves are smoother. However, the dashed lines in the figure represent that the co-prime MIMO radars can only take a few discrete integers on the abscissa, indicating that the design of nested array MIMO radar is more flexible.

Table 2. The closed-form expression of attainable consecutive DOFs for five types of MIMO radar.

	The Total Number of Physical Radar Elements	The Closed-Form Expression of Consecutive DOFs
UCA-MIMO radar in [20]	$M' + N'$	$2(3MN - M - N) - 1$
GTNA-MIMO radar in [19]	$M_1 + M_2 + N_1 + N_2$	$2\alpha M_2(M_1 - 1) + 2\beta N_2(N_1 + 1)$
NA-MIMO radar in [19]	$M_1 + M_2 + N_1 + N_2$	$2M_2N_2(M_1 + 1)(N_1 + 1) - 1$
UGCA-MIMO radar in [24]	$M' + N'$	$8(MN)^2 - 1$
The proposed GNA-MIMO radar	$M_1 + M_2 + N_1 + N_2$	Equation (14)

$\alpha = M_1M_2 - M_1 - M_2.\beta = \alpha + 1$. M' and N' are coprime.

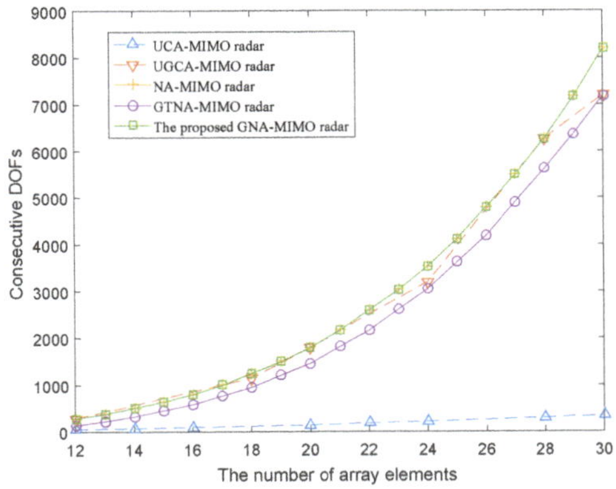

Figure 2. The attainable consecutive DOFs with given number of physical sensors.

4.2. The Mutual Coupling Leakage Rate

This subsection compares the coupling leakage rates of different MIMO radars under the optimal array element configuration. Generally, the smaller the mutual coupling leakage rate, the less influence the MIMO radar experiences in DOA estimation, resulting in more accurate DOA estimation results. The mutual coupling is built on (6) with $c_0 = 1$, $c_1 = 0.3e^{j\pi/3}$, $B = 100$, $c_l = c_1 e^{-j(l-1)\pi/8}/l$ for $2 \leq B \leq 100$ [30].

Figure 3 compares the coupling leakage rates for these five MIMO radar systems when the total number of array elements is 12 and 24. From Figure 3, it can be observed that the proposed GNA-MIMO radar exhibits a lower coupling leakage rate compared to the other MIMO radars. Additionally, it can be observed that UCA-MIMO radar exhibits the highest mutual coupling leakage rate, mainly resulting in smaller element spacing and a significant mutual coupling effect. The mutual coupling leakage rate of NA-MIMO radar is also significant, especially when there are lots of array elements. The reason is that NA-MIMO radar contains a large number of adjacent array elements, which increases as the number of array elements increases.

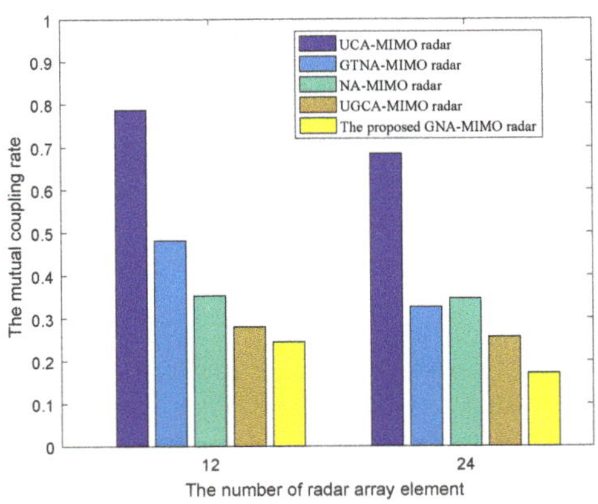

Figure 3. The coupling leakage rates for these five MIMO radar systems when the total number of array elements is 12 and 24.

4.3. The CRB Bound

The CRB bound means that the variance of an unbiased estimator can only approximate the CRB infinitely but not better than it. Thus, CRB provides a standard for the performance of an unbiased estimator. The CRB expression in this paper is derived according to the proof in the literature [22],

$$\mathrm{CRB}_{\theta} = \frac{1}{L}\left(\mathbf{F}_{\theta}^{H} \sqcap_{\mathbf{F}_s}^{\perp} \mathbf{F}_{\theta}\right)^{-1}, \tag{16}$$

where $\mathbf{F}_{\theta} = (\mathbf{R}_x{}^T \otimes \mathbf{R}_x)^{-\frac{1}{2}} \hat{\mathbf{C}}\mathbf{R}_s$, $\sqcap_{\mathbf{F}_s}^{\perp} = \mathbf{I} - \mathbf{F}_s(\mathbf{F}_s{}^H\mathbf{F}_s)^{-1}\mathbf{F}_s{}^H$, $\mathbf{F}_s = (\mathbf{R}_x{}^T \otimes \mathbf{R}_x)^{-\frac{1}{2}}\mathbf{A}_0$, $\hat{\mathbf{C}} = \hat{\mathbf{A}}^* \odot \mathbf{A} + \mathbf{A}^* \odot \hat{\mathbf{A}}$ and $\hat{\mathbf{A}} = \frac{\partial}{\partial\theta}\mathbf{A}$.

Figure 4 shows the CRB curves of five MIMO radars where the SNR ranges from -15 dB to 15 dB and snapshots $L = 100$. It can be observed from the figure that the CRB curve of the proposed GNA-MIMO radar has a smaller corresponding value than that of the other four MIMO radars. Therefore, the MIMO radar designed in this paper has the best DOA performance theoretically.

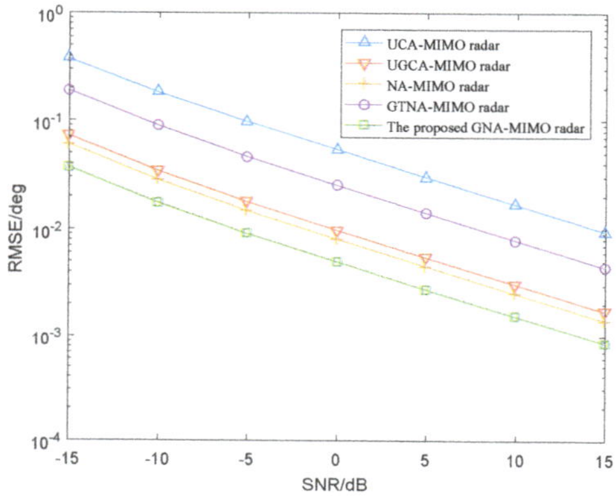

Figure 4. The CRB curves of five MIMO radars when the snapshots $L = 100$.

5. Simulation Results

In this part, we provide numerical simulation results to demonstrate the effectiveness and superiority of our proposed GNA-MIMO radar. First, we consider the optimal five MIMO radar configurations mentioned in the last section. Each of them is composed of 12 physical sensors. The mutual coupling is also built on (6) with $c_0 = 1$, $c_1 = 0.3e^{j\pi/3}$, $B = 100$ and $c_l = 0.3e^{-j(l-1)\pi/8}/l$ for $2 \leq B \leq 100$ [30]. In the experiment, the signal power of the receiver array is fixed at 16 DBm. The transmitted signal is a Gaussian signal and the noise simulation of the channel is additive Gaussian white. Subsequently, the SS-MUSIC algorithm is utilized to estimate DOA. At last, the results of the MUSIC spectrum, the root mean square error (RMSE) curves and angle resolution performance are obtained.

5.1. MUSIC Spectrum

Figure 5 shows the spatial spectrum with the SS-MUSIC algorithm for four radar systems, except for the UCA-MIMO radar in [25]. Because of the limited consecutive DOFs of the UCA-MIMO radar, effective spatial spectra cannot be obtained. Assuming that there are $K = 31$ incoherent targets in the far field evenly distributed from $-30°$ to $30°$. SNR = 3 dB; the number of snapshots $L = 100$; the search range is from $-31°$ to $31°$ and the step size is $0.01°$. It can be seen from the figure that the MIMO radar designed in this paper can estimate incoherent targets far beyond the total number of physical array sensors. Moreover, the proposed GNA-MIMO radar exhibits more precise angular resolution and sharper spectral peaks.

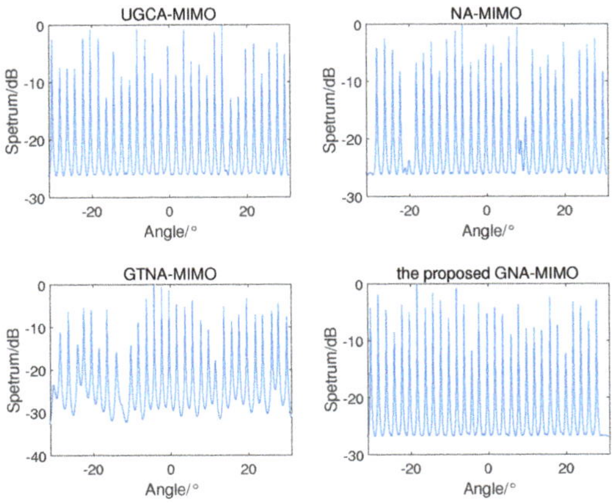

Figure 5. The spatial spectrum with SS-MUSIC when there are $K = 31$ incoherent targets.

5.2. The RMSE Performance versus SNR and Snapshots

We utilize the SS-MUSIC algorithm to research the different performances of DOA estimation with five MIMO radars. It is assumed there are $K = 6$ incoherent targets impinging on the MIMO radar from angles [10°, 20°, 30°, 40°, 50°, 60°]. The RMSE of estimated DOAs can be calculated by [20]

$$\text{RMSE} = \sqrt{\frac{1}{K\Gamma}(\sum_{k=1}^{K}\sum_{\tau=1}^{\Gamma}\theta_k - \hat{\theta}_{k,\tau})^2}, \tag{17}$$

where Γ is the number of Monte-Carlo trials, $\hat{\theta}_{k,\tau}$ denotes the τ-th trial of the k-th angle θ_k and θ_k represents the true DOA.

Figure 6 depicts the DOA estimation performance with different MIMO radars versus SNR, where the number of snapshots $L = 1000$. Figure 7 illustrates the RMSE performance versus the number of snapshots, where SNR is set as 0 dB. It can be observed that the DOA estimation results become more and more accurate with the increase in snapshots and SNR. Specifically, under the condition of given snapshots and SNR, the proposed GNA-MIMO radar outperforms other MIMO radars due to its ability to provide pretty much more consecutive DOFs and a lower mutual coupling rate.

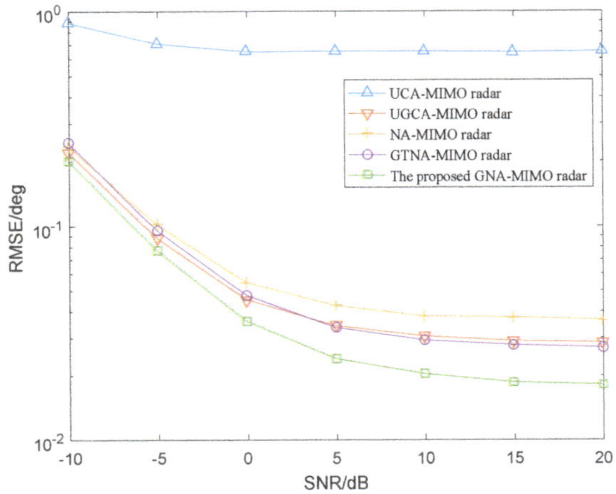

Figure 6. The RMSE performance versus SNR when $L = 1000$.

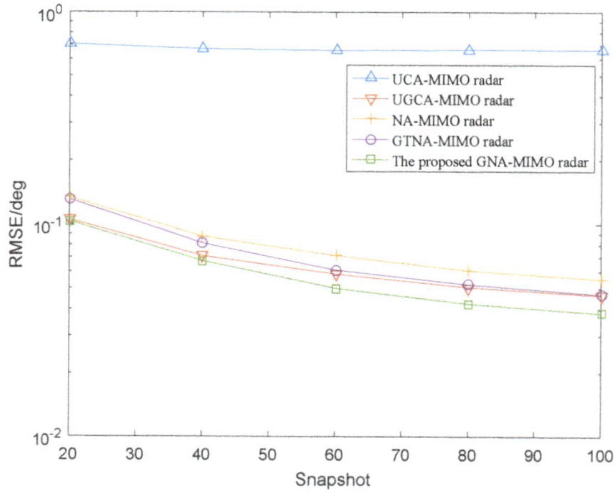

Figure 7. The RMSE performance versus the number of snapshots when SNR = 0.

5.3. Resolution Performance

Assuming that there are two close uncorrelated targets located at θ_1 and θ_2 and the estimation results of these two angles are, respectively, denoted as $\hat{\theta}_1$ and $\hat{\theta}_2$. If θ_1, θ_2, $\hat{\theta}_1$, $\hat{\theta}_2$ satisfy that $|\theta_1 - \hat{\theta}_1| < |\theta_1 - \theta_2|/2$ and $|\theta_2 - \hat{\theta}_2| < |\theta_1 - \theta_2|/2$, we consider these two targets can be successfully resolved [21]. In this section, we assume $\Delta\theta = |\theta_1 - \theta_2| = 0.5°$.

Figure 8 depicts the success rate of angle resolution with the variation of SNR when the number of snapshots is 10. Figure 9 demonstrates the success rate of angle resolution with the variation of snapshot number where SNR is −6 dB. From Figures 8 and 9, it can be observed that as the SNR and snapshot number increase, the success rate of angle resolution improves. Obviously, under the same simulation conditions the proposed GNA-MIMO radar achieves the highest success rate of angle resolution.

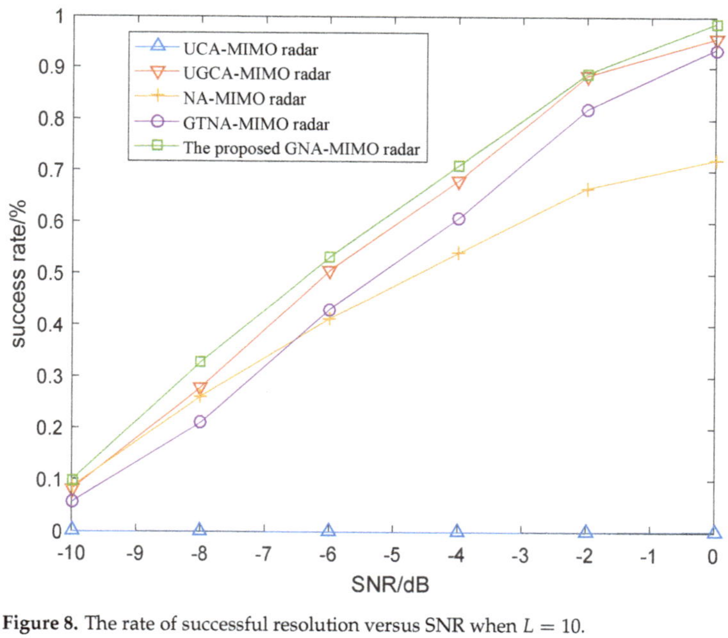

Figure 8. The rate of successful resolution versus SNR when $L = 10$.

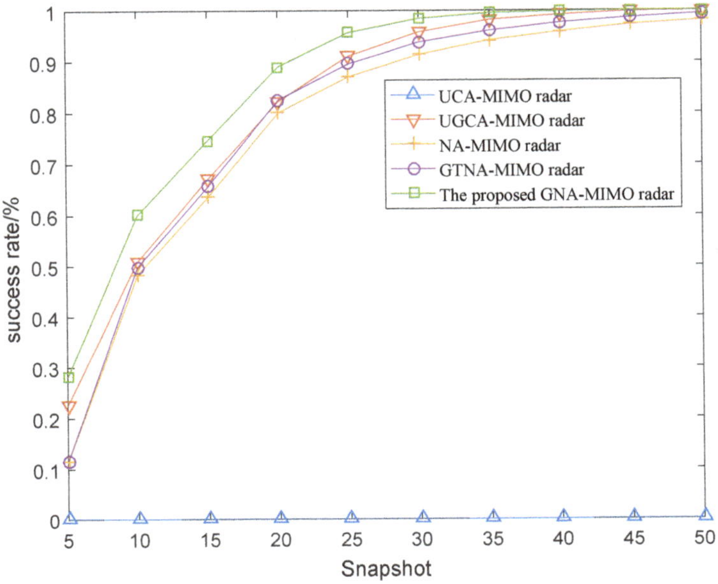

Figure 9. The rate of successful resolution versus the number of snapshots when $SNR = -2$.

6. Conclusions

This paper introduces a GNA-MIMO radar configuration with improved DOA performance. In the meantime, the closed-form expressions for the position of physical sensors and the attainable consecutive DOFs are derived. Simulation results verify the advantages of the proposed MIMO radar in DOA estimation performance and angle resolution using the SS-MUSIC algorithm. Moreover, our findings emphasize the importance of considering mutual coupling in MIMO radar design and highlight the efficacy of the proposed GNA-MIMO radar system in overcoming these challenges. This research opens up new possibilities for improving the performance of array signal processing techniques in practical applications. Furthermore, it is possible to substitute the receiver array with other sparse arrays to further increase DOFs while reducing the mutual coupling leakage rate.

Author Contributions: Conceptualization, Z.Y. and H.H.; methodology, Z.Y.; software, Z.Y. and H.H.; validation, Z.Y., H.H. and Y.C.; formal analysis, Z.Y. and H.H.; investigation, Z.B.; resources, Z.Y.; data curation, Z.Y.; writing—original draft preparation, Z.Y.; writing—review and editing, Y.C.; visualization, Z.Y. and Z.B.; supervision, Z.Y.; project administration, Z.Y.; funding acquisition, Z.Y. All authors have read and agreed to the published version of the manuscript.

Funding: This research was funded by China NSF Grants (61971217, 61971218, 61631020), Jiangsu NSF Grant (BK20200444) and National Key Research and Development Project Grant (2020YFB1807602).

Institutional Review Board Statement: Not applicable.

Informed Consent Statement: Not applicable.

Data Availability Statement: Data are contained within the article.

Conflicts of Interest: The authors declare no conflicts of interest.

References

1. Zhang, X.; Cao, R. *Direction of Arrival Estimation: Introduction*; Wiley: London, UK, 2017.
2. Qin, S.; Zhang, Y.D.; Amin, M.G.; Zoubir, A. Generalized coprime sampling of Toeplitz matrices. In Proceedings of the 2016 IEEE International Conference on Acoustics, Speech and Signal Processing (ICASSP), Shanghai, China, 20–25 March 2016; pp. 4468–4472.

3. Zhang, X.; Lai, X.; Zheng, W.; Wang, Y. Sparse Array Design for DOA Estimation of Non-Circular Signals: Reduced Co-Array Redundancy and Increased DOF. *IEEE Sens. J.* **2021**, *21*, 27928–27937. [CrossRef]
4. Lai, X.; Zhang, X.; Zheng, W.; Ma, P. Spatially Smoothed Tensor-Based Method for Bistatic Co-Prime MIMO Radar With Hole-Free Sum-Difference Co-Array. *IEEE Trans. Veh. Technol.* **2022**, *71*, 3889–3899. [CrossRef]
5. Bliss, D.W.; Forsythe, K.W. Multiple-input multiple-output (MIMO) radar and imaging: Degrees of freedom and resolution. In Proceedings of the The Thrity-Seventh Asilomar Conference on Signals, Systems Computers, Pacific Grove, CA, USA, 9–12 November 2003; Volume 1, pp. 54–59
6. Cao, Y.; Zhang, Z.; Dai, F.; Xie, R. Fast DOA estimation for monostatic MIMO radar with arbitrary array configurations. In Proceedings of the 2011 IEEE CIE International Conference on Radar, Chengdu, China, 24–27 October 2011; Volume 1, pp. 959–962.
7. Yang, M.; Sun, L.; Yuan, X.; Chen, B. A New Nested MIMO Array With Increased Degrees of Freedom and Hole-Free Difference Coarray. *IEEE Signal Process. Lett.* **2018**, *25*, 40–44. [CrossRef]
8. Wandale, S.; Ichige, K. A Nested Array Geometry with Enhanced Degrees of Freedom and Hole-Free Difference Coarray. In Proceedings of the 2021 29th European Signal Processing Conference (EUSIPCO), Dublin, Ireland, 23–27 August 2021; pp. 1905–1909.
9. Chen, C.Y.; Vaidyanathan, P.P. Minimum redundancy MIMO radars. In Proceedings of the 2008 IEEE International Symposium on Circuits and Systems (ISCAS), Seattle, WA, USA, 18–21 May 2008; pp. 45–48.
10. Liao, G.; Jin, M.; Li, J. A two-step approach to construct minimum redundancy MIMO radars. In Proceedings of the 2009 International Radar Conference "Surveillance for a Safer World" (RADAR 2009), Bordeaux, France, 12–16 October 2009; pp. 1–4.
11. Dong, J.; Li, Q.; Guo, W. A Combinatorial Method for Antenna Array Design in Minimum Redundancy MIMO Radars. *IEEE Antennas Wirel. Propag. Lett.* **2009**, *8*, 1150–1153. [CrossRef]
12. Dong, J.; Shi, R.; Lei, W.; Guo, Y. Minimum Redundancy MIMO Array Synthesis by means of Cyclic Difference Sets. *Int. J. Antennas Propag.* **2013**, *2013*, 323521. [CrossRef]
13. Pal, P.; Vaidyanathan, P.P. Nested Arrays: A Novel Approach to Array Processing With Enhanced Degrees of Freedom. *IEEE Trans. Signal Process.* **2010**, *58*, 4167–4181. [CrossRef]
14. Vaidyanathan, P.P.; Pal, P. Sparse Sensing With Co-Prime Samplers and Arrays. *IEEE Trans. Signal Process.* **2011**, *59*, 573–586. [CrossRef]
15. Shi, J.; Hu, G.; Zhang, X.; Zhou, H. Generalized Nested Array: Optimization for Degrees of Freedom and Mutual Coupling. *IEEE Commun. Lett.* **2018**, *22*, 1208–1211. [CrossRef]
16. Qin, S.; Zhang, Y.D.; Amin, M.G. Generalized coprime array configurations. In Proceedings of the 2014 IEEE 8th Sensor Array and Multichannel Signal Processing Workshop (SAM), A Coruna, Spain, 22–25 June 2014; pp. 529–532.
17. Li, J.; Zhang, X. Direction of Arrival Estimation of Quasi-Stationary Signals Using Unfolded Coprime Array. *IEEE Access* **2017**, *5*, 6538–6545. [CrossRef]
18. Qin, S.; Zhang, Y.D.; Amin, M.G. DOA estimation of mixed coherent and uncorrelated signals exploiting a nested MIMO system. In Proceedings of the 2014 IEEE Benjamin Franklin Symposium on Microwave and Antenna Sub-systems for Radar, Telecommunications, and Biomedical Applications (BenMAS), Philadelphia, PA, USA, 26 September 2014; pp. 1–3.
19. Qin, S.; Zhang, Y.D.; Amin, M.G. DOA estimation of mixed coherent and uncorrelated targets exploiting coprime MIMO radar. *Digit. Signal Process.* **2017**, *61*, 26–34. [CrossRef]
20. Zheng, W.; Zhang, X.; Shi, J. Sparse Extension Array Geometry for DOA Estimation With Nested MIMO Radar. *IEEE Access* **2017**, *5*, 9580–9586. [CrossRef]
21. Shi, J.; Hu, G.; Zhang, X.; Sun, F.; Zheng, W.; Xiao, Y. Generalized Co-Prime MIMO Radar for DOA Estimation With Enhanced Degrees of Freedom. *IEEE Sens. J.* **2018**, *18*, 1203–1212. [CrossRef]
22. BouDaher, E.; Ahmad, F.; Amin, M.G. Sparsity-Based Direction Finding of Coherent and Uncorrelated Targets Using Active Nonuniform Arrays. *IEEE Signal Process. Lett.* **2015**, *22*, 1628–1632. [CrossRef]
23. Zhang, Y.; Hu, G.; Zhou, H.; Zhu, M.; Zhang, F. DOA Estimation of a Novel Generalized Nested MIMO Radar with High Degrees of Freedom and Hole-Free Difference Coarray. *Math. Probl. Eng.* **2021**, *2021*, 1–9. [CrossRef]
24. Shaikh, A.H.; Dang, X.; Ahmed, T.; Huang, D. New Transmit-Receive Array Configurations for the MIMO Radar With Enhanced Degrees of Freedom. *IEEE Commun. Lett.* **2020**, *24*, 1534–1538. [CrossRef]
25. Wei, Z. DOA Estimation of Single-Base Unfolded Co-prime Array MIMO Radar. *J. Nanjing Univ. Posts Telecommun. Nat. Sci. Ed.* **2019**, *39*, 8.
26. Wei, Z.; Qiang, W.; Jun, T.; Wei, Z.; Yingjie, P. Dimensionality Reduction Estimation of DOD and DOA for Dual-Base Unfolded Co-prime Array MIMO Radar. *J. Nanjing Univ. Posts Telecommun. Nat. Sci. Ed.* **2020**, *40*, 7.
27. Yulei, Z.; Guoping, H.; Hao, Z.; Shijie, Y.; Feilong, Z. DOA Estimation of High-Degree-of-Freedom Low-Mutual-Coupling Generalized Second-Order Nested MIMO Radar. *Syst. Eng. Electron.* **2021**, *43*, 2819–2827.
28. Hao, H.; Lai, X.; Han, S.; Zhang, X. Unfolded augmented co-prime array for MIMO radar: Low mutual coupling and high degree of freedom perspective. *J. Appl. Sci.* **2023**, *41*, 911–925.
29. Li, J.; He, Y.; Ma, P.; Zhang, X.; Wu, Q. Direction of Arrival Estimation Using Sparse Nested Arrays With Coprime Displacement. *IEEE Sens. J.* **2021**, *21*, 5282–5291. [CrossRef]

30. Li, J.; Stoica, P., MIMO Radar Diversity Means Superiority. In *MIMO Radar Signal Processing*; IEEE: Piscataway, NJ, USA, 2009; pp. 1–64.
31. Huang, Y.; Liao, G.; Li, J.; Wang, H. Sum and difference coarray based MIMO radar array optimization with its application for DOA estimation. *Multidimens. Syst. Signal Process.* **2017**, *28*, 1–20. [CrossRef]

sensors

MDPI

Review

Role of Radio Telescopes in Space Debris Monitoring: Current Insights and Future Directions

Bhaskar Ahuja [1,2], Luca Gentile [2], Ajeet Kumar [2,*] and Marco Martorella [2,3]

1 Department of Physics, University of Trento, Via Sommarive 14, 38123 Trento, Italy; bhaskar.ahuja@unitn.it
2 RaSS National Laboratory, National Inter-University Consortium for Telecommunications (CNIT), 56124 Pisa, Italy; luca.gentile@cnit.it (L.G.); m.martorella@bham.ac.uk (M.M.)
3 Microwave Integrated Systems Laboratory (MISL), University of Birmingham, Birmingham B15 2TT, UK
* Correspondence: ajeet.kumar@cnit.it

Abstract: The growing population of space debris poses significant risks to operational satellites and future space missions, necessitating innovative and efficient tracking solutions. Ground-based radar for space surveillance has been a central area of research since the early Space Age, with recent advancements emphasizing the use of bistatic radar systems that incorporate sensitive radio telescopes as receivers. This approach offers a cost-effective and scalable solution for monitoring space debris. Preliminary observations demonstrated the viability of employing radio telescopes in bistatic configurations for effective debris tracking. This review provides a comprehensive analysis of experiments utilizing radio telescopes as bistatic receivers, highlighting key advancements, challenges, and potential applications in space surveillance systems. By detailing the progress in this field, this study underscores the critical role of bistatic radar systems in mitigating the growing threat of space debris.

Keywords: radio telescope; bistatic radar; debris tracking; space surveillance and tracking; passive radar

1. Introduction

1.1. Background and Motivation

Space debris presents several significant risks to space operations and long-term sustainability of space activities [1]. Collision hazards are a primary concern, as debris can damage or destroy operational satellites, spacecraft, and the International Space Station (ISS). The risk of Kessler Syndrome, in which debris collisions generate more debris, threatens the viability of certain orbital regions, particularly the Low Earth Orbit (LEO) [2]. The presence of space debris also increases the operational costs for space agencies and satellite operators, necessitating investment in tracking systems, collision avoidance maneuvers, and protective shielding. Ultimately, the increasing quantity of debris threatens the long-term sustainability of space activities, presenting a complex challenge for the global space community. As part of Space Situational Awareness (SSA) [3], Space Surveillance and Tracking (SST) focuses on technologies such as radar, optical telescopes, and advanced data analysis algorithms to identify and forecast the paths of active satellites and space debris with precision [4]. A global network of surveillance sites is continuously being developed to ensure uninterrupted space monitoring. The United States maintains a comprehensive catalog through its Space Surveillance Network (SSN), utilizing both optical and radar sensors [5,6]. Based on the data collected by these sensors, Two-Line Element (TLE) sets

https://doi.org/10.3390/s25092900

are generated and made freely available to promote international cooperation, enabling orbit propagation using the standard SGP4 algorithm [7].

Radar has been employed in space surveillance studies since the early years of the Space Age because of its key benefit of operating efficiently under all weather conditions without the need for sunlight, which optical sensors do [8]. In addition to traditional radar systems, complementary techniques such as phased array radars and LIDAR (Light Detection and Ranging) have been employed for tracking smaller space debris due to their enhanced accuracy and rapid beam steering [9,10]. However, the high cost of phased arrays and the technical challenges associated with high-power lasers in LIDAR systems limit their scalability [11,12]. Thus, research is increasingly focusing on developing innovative and cost-effective solutions for space surveillance by leveraging existing infrastructure and passive sensing techniques.

In particular, systems that employ a sensitive radio telescope as a bistatic receiver operating in beam-park mode have gained interest, due to their potential for wide-area coverage with reduced infrastructure requirements [13]. Radio telescopes, originally designed for observing faint cosmic signals [14], have demonstrated their capability as receivers for radar applications. Researchers have been investigating the use of radio telescopes as radar receivers for space surveillance applications, and several experiments have been performed with configurations of interest. A few of these experiments, with a special focus on the LEO region, were reported in [15]; however, an extensive review is still missing. This article comprehensively examines debris monitoring experiments conducted using radio telescopes in the past, provides a detailed analysis, and elucidates the progress made to date in this field. Figure 1 shows the locations of radio telescopes around the globe that have been utilized in past space monitoring campaigns. Several countries, including Australia and South Africa, have contributed to such efforts; however, Europe remains the primary contributor, particularly in advancing innovative radio-telescope-based approaches. Furthermore, the classification of systems into active and passive types, based on their transmitter configuration, is explained in the following section.

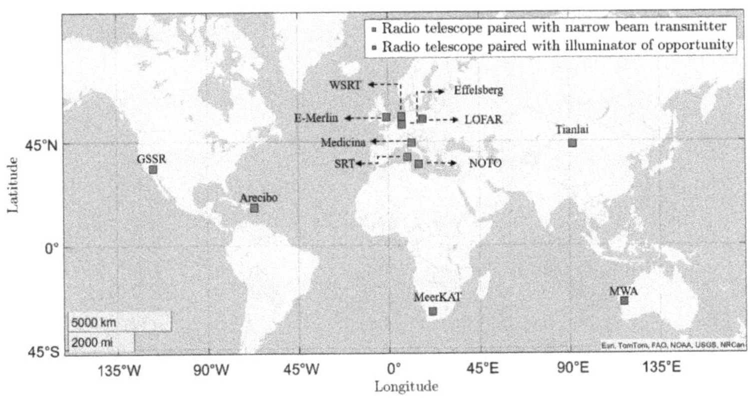

Figure 1. Locations of radio telescopes around the world involved in past space debris monitoring campaigns. Blue squares represent systems used with active narrow beam transmitters, while red square indicate passive systems that utilize illuminators of opportunity as transmitters (generated using MATLAB R2024a, www.mathworks.com; accessed on 6 March 2025).

1.2. Structure of the Review

The aim of this section is to clarify the boundaries of the review and explain how the content is organized. Firstly, the suitability of radio telescopes for space debris monitoring, along with their advantages, is discussed in Section 2. Secondly, based on the available literature, the radio telescope experiments have been categorized into two main state-of-the-art configurations, classified by the type of transmitter involved.

- Active systems are those in which the radio telescope acts as a receiver in conjunction with a separate high-power narrow-beam radar transmitter. These systems emit controlled signals and analyze the reflections from space objects. However, the narrow beams of the radar and receiver limit the coverage area. This is consistent with the operation of most space-surveillance radar systems that use single narrow beams or fence configurations. One example of this kind of system is the Effelsberg radio telescope with Tracking and Imaging Radar (TIRA) as the transmitter [16]. Figure 2a illustrates an active bistatic configuration of the TIRA–Effelsberg system, where a narrow beam transmitter TIRA illuminates the target and the Effelsberg radio telescope captures the reflected signal.
- Passive bistatic radar systems rely on existing sources of illumination, such as FM and DVB signals. Unlike active radar systems, passive radar does not require a dedicated transmitter, making it the most cost-effective alternative. In this setup, single or distributed arrays of radio telescopes are paired with broadband transmitters located at significant distances. The wider field of view of the receiver, combined with the broad broadcast coverage of the transmitter, results in a significantly larger surveillance volume. Although this setup decreases the power concentration on space objects, it offers the benefit of a longer detection time for space debris, allowing precise orbital calculations without requiring previous data. Moreover, a distributed ground broadcast network allows several simultaneous sources of illumination. One such example is the Murchison Widefield Array (MWA) in Australia [17]. Figure 2b illustrates a passive radar configuration, where MWA functions as a receiving system capturing signals reflected from a space object, using non-cooperative transmitters (illuminators of opportunity).

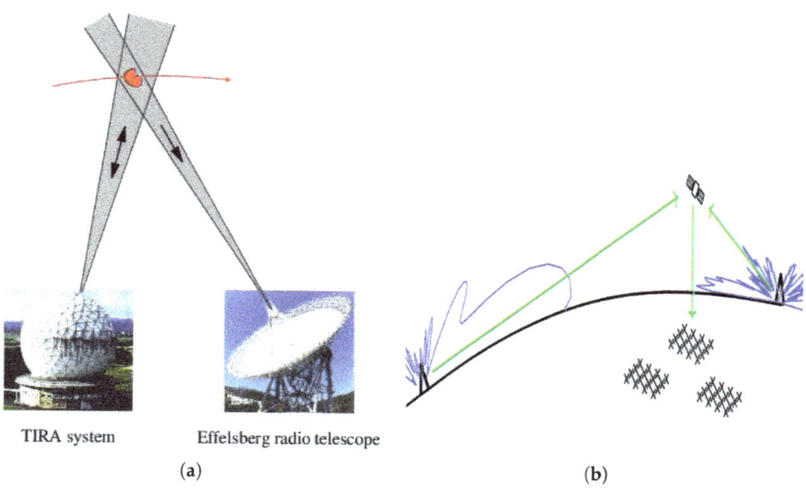

Figure 2. Types of configurations based on the nature of the transmitter used: (**a**) An example of a narrow-beam active transmitter system: TIRA used as the transmitter and Effelsberg as the receiver, reproduced from [16]. (**b**) An example of a passive configuration using illuminators of opportunity: Murchison Widefield Array (MWA) operating as a receiver with external transmitters, reproduced from [17].

The structure of the following sections reflects this division, with separate discussions on experiments with active transmitters and passive setups using illuminators of opportunity. Section 3 reviews experiments conducted with narrow beam transmitters, and Section 4 then discusses experiments carried out involving illuminators of opportunity. Figure 3 illustrates the structure of the review with classification of radio telescope experiments based on the type of transmitter used and lists the notable telescopes involved in experiments conducted under each configuration. Lastly, the conclusions and discussion are presented in the final section of the article. In summary, this article discusses the outcomes of the experiments conducted and presents a broad overview of the current state of research, along with a discussion of future scope.

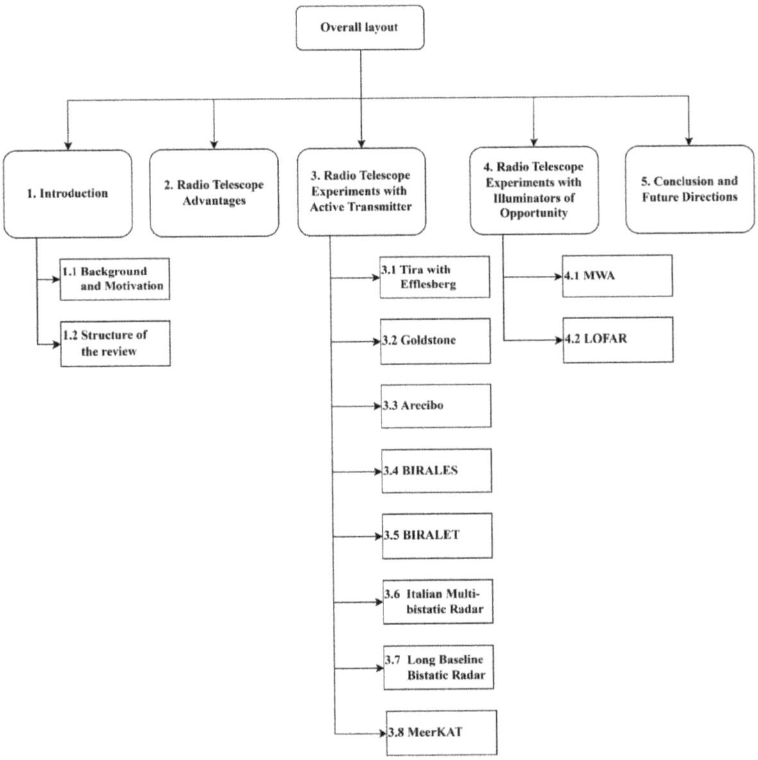

Figure 3. This flowchart presents the overall layout of the review article, including classification of radio telescope experiments based on the type of transmitter used and the radio telescopes involved in the past experiments.

2. Why Use Radio Telescopes?

One of the main reasons behind using radio telescopes as receivers is their capability to detect small debris that could not be observed by traditional monostatic radars. This can be understood by the well-known radar Equation in (1) ([18] pp. 1.11–1.13).

$$SNR = \frac{P_t G_t G_r \lambda^2 \sigma}{(4\pi)^2 R_t^2 R_r^2 L k T_{sys} B} \tag{1}$$

where P_t denotes transmitted power. G_t, G_r are the transmitter and receiver gains, λ is the wavelength, σ is the radar cross-section (RCS) of the debris. R_t, R_r are the range from the target to the transmitter and receiver, respectively, k is Boltzmann's constant, T_{sys} represents the system noise temperature, B represents the bandwidth of the system, and

SNR represents the signal-to-noise ratio. The benefits of using a radio telescope as a bistatic receiver are summarized as follows:

1. The high sensitivity of radio telescopes arises from their large collecting areas. According to antenna theory, receiver antenna gain (G_r) in Equation (1) can be expressed as

$$G_r = \frac{4\pi A_{eff}}{\lambda^2} \quad (2)$$

where A_{eff} denotes the effective reception area. The larger the effective area, the larger the signal-to-noise ratio. Hence, the large aperture size of the radio telescopes significantly enhances their detection range.
2. The lower system noise temperature (T_{sys}) of radio telescopes is another key factor contributing to their higher SNR, as it is inversely proportional. This improved sensitivity is achieved by cryogenically cooling the receiver front-end, including the low-noise amplifiers, to temperatures as low as 4 Kelvin using liquid helium. Such cryogenic cooling significantly reduces thermal noise and enhances the overall sensitivity of the system [19].
3. In a bistatic configuration, separating the transmitter and receiver enhances angular diversity, minimizing blind spots and improving detection capabilities. The total bistatic range, given by $R_b = R_t + R_r$, provides greater flexibility in observation geometry. This configuration complements traditional monostatic radar systems by enhancing coverage and redundancy [20].
4. Radio telescopes can also contribute to debris characterization by analyzing radar cross-sections σ at multiple scattering angles, providing detailed information about the size, shape, and material composition of debris, as σ is a function of $f(\theta_{inc}, \theta_{sca}, \phi, \lambda)$, where θ_{inc} is the incident angle, θ_{sca} is the scattering angle, and ϕ is the polarization. This detailed profiling is vital for understanding the physical properties of debris and assessing its potential threats [21].
5. Interferometric capabilities from modern phased array radio telescopes such as the Square Kilometre Array (SKA) enable high-resolution imaging of space debris, allowing precise visualization of debris clouds and their spatial distribution.
6. Using these radio telescopes in a multistatic configuration becomes beneficial, as the systems already work in synchronization for Very Long Baseline Interferometry (VLBI) or similar projects.
7. Another benefit of using a radio telescope is to utilize the existing infrastructure, primarily designed for other purposes, which makes the overall system cost-effective.

These features, coupled with the ability to leverage the existing infrastructure, make radio telescopes indispensable tools for enhancing SSA.

3. Radio Telescope Measurements with Active Transmitters

Previous investigations have shown that several available facilities around the globe participated in collaborative space debris monitoring, based on the first architecture discussed in Section 1.2. These measurement campaigns have included Germany's Effelsberg Radio Telescope in collaboration with TIRA since 1996 [8], which is described in detail in Section 3.1. Another validation experiment was performed in the late 90s using Goldstone bistatic Radar in the USA [22], which is discussed in Section 3.2. The Arecibo Radio telescope, once the largest telescope in the world, in Puerto Rico, reported a few debris observations during the early 1990s. Sadly, this architecture has been dismantled due to collapse in 2020, as discussed in Section 3.3 [23]. Moreover, an initial stage space debris observational test was conducted in 2007 using the Medicina–Evpatoria bistatic radar system, where the system consisted of the Evpatoria RT-70 radar antenna in Ukraine as the transmitter and the Medicina VLBI 32 m dish in Italy as the receiver [24]. Unfortunately, due to the ongoing war in Ukraine, the Evpatoria antenna is no longer available for collaboration with the Western world.

Moreover, Italy has two active bistatic radars based on radio telescopes. One of them is the BIRALES system, whose acronym stands for BI-static RAdar for LEO Survey, located in Medicina, Bologna [25]. The other is the BIRALET system, which is an acronym for BI-static RAdar for LEO Tracking, and its receiver is the Sardinia Radio Telescope (SRT) located in Sardinia [26]. A detailed review and the results from the BIRALES and BIRALET experimental campaigns are discussed in Sections 3.4 and 3.5, respectively. Our group in Pisa, Italy, has also contributed to this field by investigating the performance metrics and operational benefits of multibistatic configurations using three Italian radio telescopes as receivers [27–29], and it is discussed in detail in Section 3.6. In the last few years, experiments have been conducted by NATO Science and Technology Organization (STO) members for Geostationary Earth Orbit (GEO) monitoring [30], and it is reported in Section 3.7. A relevant additional investigation was carried out with SKA in South Africa, specifically utilizing the MeerKAT radar system, which is also explored in detail in Section 3.8.

Recent studies have further advanced the application of radio astronomy to space debris tracking. For instance, Tianlai radio arrays have been explored as part of a bistatic radar system for LEO debris detection [31]. Additionally, ref. [32] demonstrated that passive RF observations using collocated antennas can refine orbit determination and reduce TLE-based errors to below 100 m and 0.2 m/s, as validated by the Canadian Space Agency. However, these specific findings are beyond the scope of this article and have not been reported here.

3.1. *TIRA with the Effelsberg Telescope*

The Fraunhofer Institute for High-Frequency Physics and Radar Techniques (FHR), Germany, operates one of the largest radars in Europe, the Tracking and Imaging Radar (TIRA). It is capable of operating in both the L-band (1.33 GHz) as a narrow-band monopulse tracking radar and the Ku-band (16.7 GHz) as a high-range-resolution imaging radar [33]. In 1996, a space debris observation experiment was conducted using a bistatic radar formed with the TIRA system as the radar transmitter and the Effelsberg radio telescope as the radar receiver [8]. The Effelsberg radio telescope has a 100 m dish antenna and is operated by the Max Planck Institute for Radio Astronomy in Germany. Several experiments using TIRA as the transmitter/receiver and the Effelsberg radio telescope as a secondary receiver to lower the detection threshold have been performed since 1996 [34–36]. Figure 2a illustrates the geometry of the experiment, showing the TIRA transmitter and the Effelsberg radio telescope as the receiver.

The initial COoperative BEAM-park mode (COBEAM) experiment was conducted over a 24-h period and successfully collected observational data on objects as small as 0.9 cm at a range of 1000 km [34]. Figure 4a illustrates the detection rate (detections per hour) against altitude (at 30 km intervals) for the TIRA L-band radar and Effelsberg. Effelsberg achieved a notably higher detection rate owing to its higher sensitivity, even though its beam width was only one-third that of the TIRA L-band beam. Figure 4b displays the detection rates compared with range rates over the −1 to 1.8 km/s limits, segmented into 100 m/s increments. The distribution peaks at 0.1 km/s for data from TIRA, whereas the Effelsberg data predominate around 1 km/s. Following the first promising results of the COBEAM experiment, researchers have refined estimation and calibration procedures to further enhance system performance [37]. The authors in [38] showed a moderate improvement in the detection capability after revising the multidimensional maximum likelihood estimation for determining object trajectories and RCS between beam park campaigns in 2008 and 2009. Ongoing TIRA upgrades, including Ka-band polarimetric imaging and L-band tracking enhancements, will significantly boost space surveillance and support future bistatic measurement campaigns [33]. Expanding on the successful bistatic radar experiments involving TIRA-Effelsberg, which improved the detection of space debris in low Earth orbit, the German Experimental Space Surveillance and Tracking Radar (GESTRA) is set to further enhance these capabilities. Its phased-array design inherently supports bistatic and multistatic configurations, leading to improved detection accuracy and estimation performance [9].

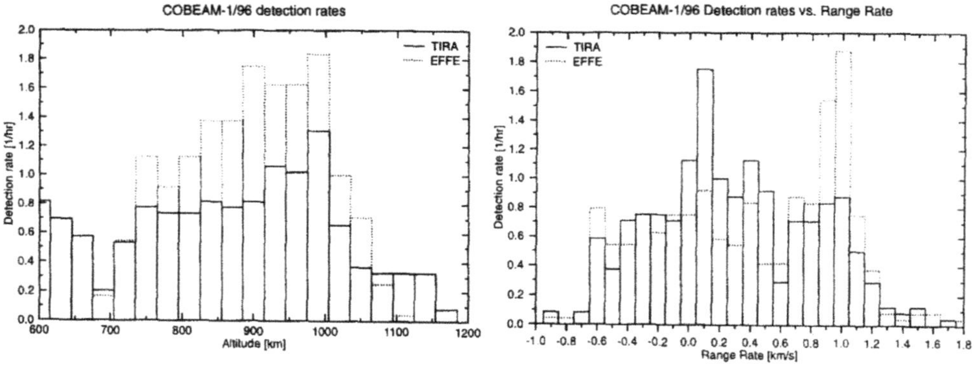

(a) (b)

Figure 4. Results from the first COBEAM experiment performed in 1996: (**a**) Detection rate (detections per hour) versus altitude, shown in 30 km bins, comparing performance of the TIRA L-band radar and the Effelsberg radio telescope. Effelsberg demonstrates significantly higher detection rates across all altitude ranges due to its superior sensitivity. (**b**) Detection rate versus range rate in 100 m/s bins over the interval −1 to 1.8 km/s. The TIRA detections peak around 0.1 km/s, indicating more detections of near-stationary or slowly moving objects, while Effelsberg shows a dominant detection rate near 1 km/s, corresponding to faster-moving targets in its field of view [34]. Reprinted from Advances in Space Research, Copyright 1999, with permission from Elsevier.

3.2. Goldstone Orbital Debris Radar

The Goldstone Orbital Debris Radar operated by NASA, is a highly capable system used for monitoring the orbital debris environment. It has supported multiple space debris observation campaigns [39]. Between October 1994 and May 1997, the Goldstone Solar System Radar, together with a 35-m parabolic antenna acting as the receiver, operated in beam-park mode. During this period, approximately 110 h of observations were conducted, resulting in the detection of 3476 objects [22]. Figure 5 presents a summary of this measurement campaign, where the minimum radar cross-section (RCS) of the detected objects is plotted against their corresponding altitudes. A particularly noteworthy finding from this study was the unexpected identification of a concentration of debris around 2900 km in altitude, a group not previously observed by other radar systems. Additionally, very few objects were detected below 300 km, which the authors attributed to the geometric limitations of the intersecting radar beam [22]. In a related study, Stokley et al. [40] compared debris flux measurements from three radar systems: the Goldstone Radar, Haystack Imaging Radar, and Haystack Auxiliary Radar (HAX) before the year 1998, during 2001, and after 2003, the last solar maximum. Furthermore, the findings from the measurement campaigns conducted between 2016 and 2017 are reported here [41]. Thanks to several upgrades in bandwidth and the integration of dual-polarization capabilities over the last few years, the detection efficiency of the system has improved significantly, enabling the detection of debris as small as 3 mm.

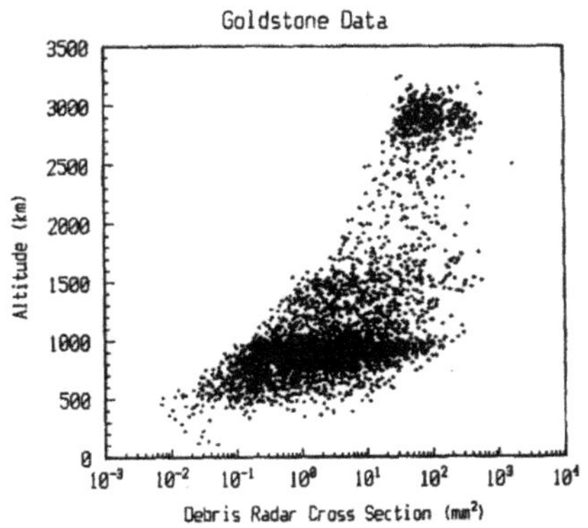

Figure 5. Radar cross-section (RCS) versus altitude of targets detected by the Goldstone bistatic radar during a 3-year period (1994–1997). The plot highlights the distribution of object sizes and altitudes, reflecting the radar's ability to detect a range of space objects across different orbital regions [22]. Reprinted from Advances in Space Research, Copyright 1999, with permission from Elsevier.

3.3. Arecibo Observatory

The Arecibo Observatory was once among the world's largest and most sensitive radio telescopes. During a debris radar experiment conducted from late June to early July 1989, the Arecibo Observatory collected approximately 19 h of data and detected 90 objects ranging in size from 5 mm to 20 mm in diameter [23]. The experiment utilized the Arecibo radar as the transmitting antenna, while a 3.5 m reflector auxiliary antenna, positioned near the main observatory, served as the receiving antenna. These antennas were separated by a 10 km baseline and were aligned to observe a common point at an altitude of 575 km, and to detect the smallest objects, ranging from 0.5 to 2 cm, by comparing the observed results with a predictive model. The comparison results from the experiment are shown in Figure 6, where the detected objects are plotted against their diameter. Unfortunately, the recent structural failure of the Arecibo dish has resulted in the loss of these significant capabilities. However, the proposal for a Next Generation Arecibo Telescope (NGAT) holds promise for improved radar performance compared with the already substantial legacy system. This enhanced sensitivity could be leveraged to address the knowledge gaps in the orbital debris environment within the size range of 1 mm to 1 cm in low Earth orbit, while simultaneously obtaining measurements across an underrepresented range of orbital inclinations [42].

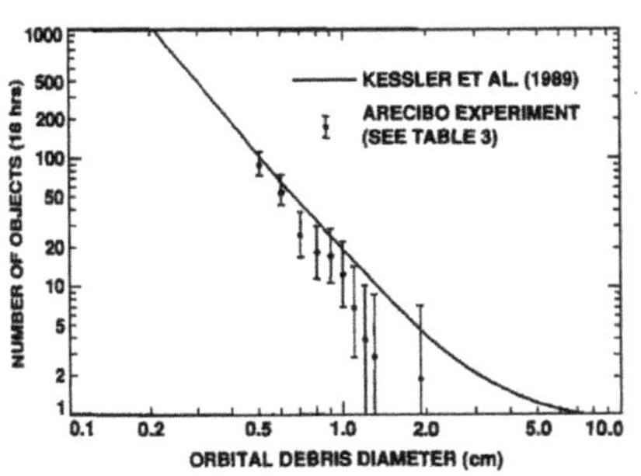

Figure 6. Comparison results from the Arecibo debris radar experiment (June–July 1989), showing detected objects plotted against diameter. Around 19 h of data collection led to the detection of 90 objects between 5 mm and 20 mm [23]. Reprinted from Geophysical Research Letters, Copyright 1992, with permission from John Wiley and Sons.

3.4. BIRALES

Both the BIRALES and BIRALET systems utilize a common transmitter, the Radio Frequency Transmitter (TRF), which consists of a 7 m dish wheel-and-track steerable parabolic antenna. The transmitter is situated in the Italian Joint Test Range within the region of Cagliari, Sardinia, Italy. TRF is capable of generating a peak power of 10 kW and operates with a 410–415 MHz bandwidth [43]. The receiving antenna is part of the Northern Cross radio telescope, which is currently one of the largest UHF-capable antennas worldwide [44], as shown in Figure 7. The components of the BIRALES receiver include eight parabolic cylindrical antennas located along the north–south (N-S) arm with a combined collection area of approximately 1400 square meters, enabling the detection of objects as small as 10 cm at a slant range of 2000 km [45]. BIRALES has achieved notable success in tracking debris owing to its higher sensitivity and strategic location in central Italy. The current 32-receiver multibeam configuration of BIRALES allows the angular track of the space object to be calculated to complement the bistatic range and Doppler measurements [46,47]. This innovative technique outperforms conventional single-beam radar because when an object passes through the field of view of the radar, it is illuminated by radio waves, and the beam illumination sequence helps define the angular trajectory of the target, as explained in [46].

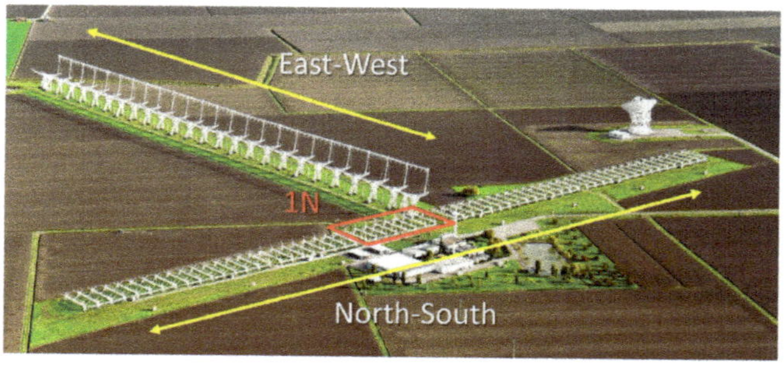

Figure 7. BIRALES receiving antenna, part of the Northern Cross radio telescope located at the Medicina Observatory in Bologna, Italy. It consists of eight parabolic cylindrical antennas arranged along the north—south arm depicted in red box, with a total collection area of approximately 1400 m^2 [44]. Reprinted from Acta Astronautica, Copyright 2020, with permission from Elsevier.

BIRALES began its measurement campaigns with initial trials in 2014, during which the bistatic radar successfully detected around 20 objects at altitudes between 300 and 850 km. A comprehensive summary of the experimental measurements performed by BIRALES in 2014 is reported in the comprehensive analysis by Muntoni et al. [15]. BIRALES is designed to simultaneously receive Continuous Wave (CW) and chirp signals from the TRF for Doppler and range measurements, respectively [48]. With the help of the multibeam approach, the time profiles of the range, Doppler, and angular position can be used to perform initial orbit determination (IOD) for unknown objects and reconstruct the track [49]. Consequently, an IOD algorithm was developed, particularly according to BIRALES system measurements, to tackle the problem of element spacing longer than half the wavelength [44]. Researchers have reported a range accuracy of 50 m and an angular accuracy of 10^{-3} deg, which were utilized by the orbit determination block to perform statistical orbit determination using measurement data from BIRALES during single- or multipass observations [50]. In 2018, BIRALES participated in tracking the re-entry of Tiangong-1 with observations conducted on 29 and 31 March. The re-entry time was predicted by propagating the state estimates, with the 31 March observation aligning well within the European Space Agency's forecasted window [51].

Another study involving BIRALES radar observation and reconstruction of the Cosmos 1408 fragmentation event, which occurred on 15 November 2021, following an anti-satellite missile test, was conducted by a team from Politecnico di Milano, the Italian Space Agency, and the Italian National Institute of Astrophysics [52]. In a recent article [53], results from the third Long March 5 B re-entry campaign, which took place from 31 October to 4 November 2022, were derived. BIRALES observations were combined with measurements from the MFDR-MR sensor, a tracking radar involved in the campaign and managed by the Italian Air Force. Orbit determination results were computed to obtain a re-entry epoch prediction, and a novel method for estimating the ballistic coefficient in a stochastic manner, which can be used in orbit determination and re-entry analysis, was also discussed by the authors. The predicted re-entry epoch was found to be consistent with the reference provided by the U.S. Space Command. A unique orbit determination pipeline was considered for the two-sensor scenario. Using the first set of observations from BIRALES, researchers calculated the initial orbit determination (IOD) with an adaptive beamforming approach [47]. Subsequently, predictions were made using an Unscented Kalman Filter (UKF) to propagate to the next observation set at the MFDR-MR tracking radar [54]. At the end of the measurements at MFDR-MR, the states were propagated again to the BIRALES sensor to obtain a second set of measurements to perform the refined orbit determination again. Table 1 presents the Keplerian parameters of the orbit determination results obtained, along with the reference parameters derived from the most recently available TLE. Note that BIRALES 1 and BIRALES 2 correspond to the first and second sets of measurements, respectively, and $a(m)$, e, i, Ω, ω, and ν represent the semimajor axis, eccentricity, inclination, longitude of the ascending node, argument of periapsis, and true anomaly, respectively.

Table 1. Keplerian parameters for the Orbit determination results obtained on 4 November from BIRALES and MFDR observations, along with parameters computed from the last available TLE [53]. Reprinted from Advances in Space Research, Copyright 2025, with permission from Elsevier.

	a [km]	e	i [deg]	Ω [deg]	ω [deg]
BIRALES 1	6530.3	0.020	41.5	77.4	179.6
MFDR-MR	6533.6	0.003	41.6	82.3	158.2
BIRALES 2	6517.4	0.004	41.6	82.1	172.6
Reference	6516.1	0.002	41.6	82.8	201.0

Overall, BIRALES demonstrates notable efficiency in detecting small objects owing to its large receiving area and real-time radar data processing backends [55]. This enables it to identify items just a few centimeters in size while significantly conserving energy. Future plans include upgrading 64 parabolic cylindrical reflectors on the N-S arm, totaling 256 receivers and over 10,000 m^2 of the collecting area, and a new transmitting antenna for observation flexibility of any kind [56]. Maximum sensitivity is achieved when all reflectors are aligned in the same direction, enabling BIRALES to detect debris a few centimeters in size in LEO. This configuration reduces the field of view from 90° × 6.6° to 5.7° × 6.6° compared with independent cylinder pointing. This mode will be effective in specific cases, such as monitoring fragment clouds after satellite explosions and to prevent collisions with operational satellites.

3.5. BIRALET

BIRALET operates in the P-band at 410–415 MHz, has a bistatic configuration composed of the same transmitting antenna used in BIRALES and SRT as the receiver, with a baseline of approximately 20 km [13]. The SRT is a 64-m fully steerable wheel-and-track antenna located near San Basilio (Cagliari, Sardinia, Italy). In the initial phase, SRT participated in the debris monitoring campaign in 2014 jointly with the Northern Cross radio telescope and was able to detect six objects, as listed in Table 2. The authors described three possible hardware configurations of the system to perform Doppler shift and range measurements [25]. The first configuration utilizes the spectrum analyzer as a backend that permits Doppler shift measurements, whereas the second configuration is based on the electronic Red Pitaya board. The final configuration uses the National Instruments USRP board as a back end. The hardware has been constantly upgraded to perform space debris monitoring, including the re-entry of the Chinese space station Tiangong-1 together with other Italian SST sensors in 2018 [57,58].

Table 2. Summary of the debris measurements by Bistatic Radar for LEO Tracking (BIRALET) in 2014, using the Sardinia Radio Telescope (SRT) as the receiver. During this campaign, six objects in LEO were successfully detected. Reproduced from [13].

Object ID	Name	Doppler (Hz)	Altitude (km)	Slant Range (km)	RCS (m^2)
22565	COSMOS 2237 (First passage)	not measured	853.6	1732	11.6
22565	COSMOS 2237 (Second passage)	−5200	852.7	2836	11.6
33320	HJ-1A (First passage)	6200	628.7	3640	1.5
33320	HJ-1A (Second passage)	4600	629.3	1832	1.5
32783	CARTOSAT 2A	5200	629.4	2056	2.3
13552	COSMOS 1408	not measured	523.7	1093	8.4
16206	COSMOS 1375	3400	984.8	2326	0.5
38047	VESSELSAT 2	5200	460.7	1664	0.3

Between 13 December 2018 and 10 October 2019, the BIRALET system was able to observe a total of 33 objects in the beam-parking mode using the dedicated back-end. The estimated and measured Doppler shifts for the detected debris between 13 December 2018 and 8 October 2019, can be found in the following literature [25,26]. Figure 8 shows a scatter plot of the RCS versus the range for the detected objects with respect to the minimum detectable size using the BIRALET system. These measurement campaigns serve as a solid foundation for monitoring space debris using radio telescopes.

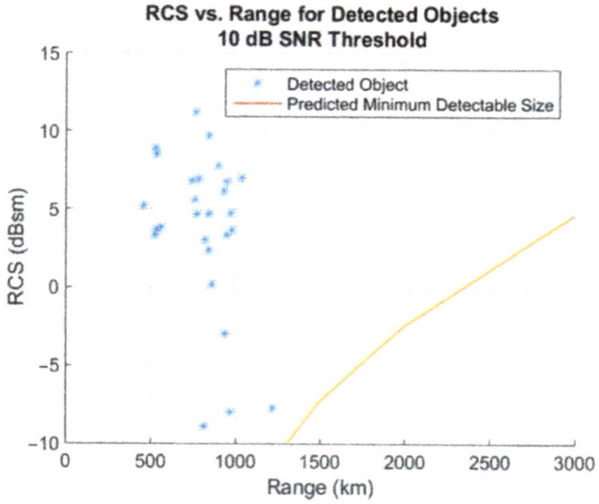

Figure 8. Scatter plot of radar cross-section (RCS) versus range for the detected objects, with reference to the minimum detectable size of the BIRALET system. Reproduced from [26].

3.6. Italian Multibistatic Radar

A multibistatic radar system consists of multiple spatially separated antennas, typically with one serving as the transmitter and several others as receivers, or vice versa. The spatial diversity provided by such configurations enhances detection, tracking, and IOD capabilities. In recent years, the Radar and Surveillance Systems National Laboratory in Pisa initiated simulation-based studies on multibistatic radar systems. Cataldo et al. [27] proposed a configuration involving three Italian radio telescopes: Medicina, NOTO, and SRT. The Medicina observatory is the same one that is used in the BIRALES experiment, while SRT is the receiver in BIRALET [48]. The third radio telescope, NOTO, located in Sicily, is operated by the Istituto di Radioastronomia di Bologna. A simulation tool was developed to evaluate system performance for SSA missions under various scenarios. To manage system complexity, the authors proposed decentralized processing with track fusion, and addressed synchronization challenges using GPS-based clocks [28]. Different signal processing techniques were implemented to extract range and Doppler measurements by combining data from multiple sensors. Consequently, IOD was performed using a multilateration technique, while orbit tracking was carried out using an Unscented Kalman Filter (UKF). Additionally, the system was designed to associate newly detected tracks or correlate them with existing ones. Figure 9a shows the target ground track on a flat Earth crossing over considered sensors, and Figure 9b presents the tracking results in Earth-Centered Earth-Fixed (ECEF) coordinates, comparing the true trajectory (blue), multilateration output (red), and UKF-based estimation (green). In a recent study [59], the same research group demonstrated that the proposed multibistatic configuration can significantly enhance tracking performance. Their findings underscore the potential of this architecture to support not only space situational awareness (SSA) but also mission planning and navigation for low-thrust transfer trajectories.

Interestingly, these three Italian radio telescopes (Medicina, NOTO, and SRT), along with the Effelsberg radio telescope in Germany, have demonstrated the capability to detect Near-Earth Objects (NEOs) in a collaborative experiment with NASA [60]. This effort was part of a European Space Agency (ESA) project aimed at defining functional requirements for a future planetary radar system and assessing the suitability of existing European infrastructure for NEO observations. In parallel, the Planetary Radio Interferometry and Doppler Experiment (PRIDE), also implemented in ESA's JUICE mission, represents a sophisticated evolution of multibistatic radar concepts [61]. PRIDE uses VLBI by coordinating globally distributed radio telescopes to derive high-precision spacecraft state vectors. Although originally developed for deep-space and planetary science applications, the PRIDE framework presents a promising approach for long-baseline, passive multibistatic tracking of high-altitude targets, such as GEO satellites. Although NEO and planetary surveillance are important applications of SSA, they lie outside the scope of this paper. Its potential for future extension to space debris monitoring is particularly relevant, as similar concepts inspire the long-baseline GEO surveillance approach discussed in the following section.

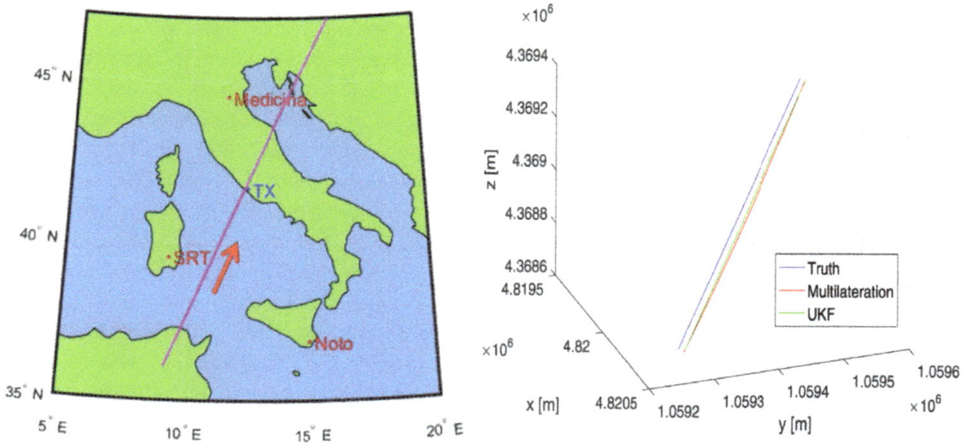

(a) (b)

Figure 9. (**a**) Ground track of the target as it passes over the Italian radio telescopes—Medicina, NOTO, and SRT. (**b**) Three-dimensional global tracking results of the target's truth path (blue), IOD-multilateration path (red), and UKF tracking path (green), demonstrating tracking accuracy and deviations [27]. Reprinted with permission from IEEE Aerospace and Electronic Systems Magazine, Copyright 2020, under license from IEEE.

3.7. Long Baseline Bistatic Radar (LBBR)

The study, conducted by the NATO STO Research Task Group (RTG), investigated the use of existing monostatic radars as transmitters in combination with large-aperture radio telescopes as bistatic receivers for GEO monitoring [30,62]. This approach aimed to enhance the sensitivity of object detection at GEO distances. Monitoring geostationary satellites poses significant challenges due to their considerable distance of approximately 36,000 km from Earth. Addressing these challenges requires high-power radars, typically in the megawatt (MW) range, as well as international collaboration to establish the necessary bistatic configuration for effective detection and tracking of objects at GEO altitudes. Table 3 lists high-power radars that are either currently used for GEO monitoring or have the capability to track GEO targets due to their substantial transmission power. Only radars that can be configured in a bistatic setup with radio telescopes are included. Other high-power phased array radars, such as the Cobra Dane phased array radar with a peak power of 15.4 MW, are not listed here. However, the full list can be found in [15].

The experiments conducted by RTG involved two transmitting radars: the Millstone Hill Radar (MHR) in the USA and TIRA in Germany. These transmitters were paired with multiple receiving radio telescopes in Europe, including SRT in Italy, Westerbork Synthesis Radio Telescope (WSRT) in the Netherlands, and multiple antennas of the e-MERLIN (enhanced MultiElement Remotely Linked Interferometer Network) array in Manchester, United Kingdom. The research team conducted several experiments between 2020 and 2021, targeting known geostationary satellites, to test the feasibility and performance of the bistatic radar setup. In such scenarios, a large-aperture radio telescope is the best option to increase the sensitivity of the system, especially when dealing with large-distance detection. Consequently, the results demonstrate successful continental baseline coherent bistatic radar collections of GEO targets. It should be noted that there was no synchronization between the transmitters and receivers, and the only available time synchronization was provided by GPS, which was presumed sufficient for these experiments.

Table 3. List of high-power radar transmitters with the capability to detect GEO targets. The table includes radar systems that are either currently used for GEO monitoring or are suitable for such applications.

Name	Country of Origin	Peak Power	Operating Frequency
TIRA	Germany	1.6 MW	1.33 GHz
Millstone Hill Radar	USA	3 MW	1.295 GHz
Haystack Radar	USA	400 kW	10 GHz
Goldstone Radar	USA	440 kW	8.56 GHz
EISCAT * UHF Radar	Sweden	2 MW	930 MHz
EISCAT Svalbard Radar (ESR)	Norway	1 MW	500 MHz

* European Incoherent Scatter Scientific Association.

The team was able to perform range–Doppler (RD) processing after performing coherent integration for a long period and then determine the target RCS. The experimental outcomes for two targets (ID 43039 and 27683) in January 2020 using the MHR-WSRT combination are displayed in Figure 10. Specifically, Figure 10a,b present the range–time intensity plot, whereas Figure 10c,d show the RD maps for the same targets. It is important to note that the origin (0, 0) of the RD maps aligns with the expected target range and Doppler measurements. The bias in the range value plots could be a result of the timing difference between the transmit and receive site clocks and the TLE error, which can reach up to 25 km in the along-track direction and 10 km in the cross-track direction for GEO objects [63]. The observed Doppler offset may have resulted from an unmatched bistatic Doppler during pulse compression, leading to a residual phase ramp across the pulses. The research team also provided measurement results from the monostatic radar TIRA and bistatic results combined with the e-MERLIN radio telescope. During this measurement campaign, three antennas (Knockin: 25 m, Pickmere: 25 m, and Cambridge: 32 m) of the e-MERLIN array simultaneously received the TIRA signal at 1.33 GHz in both polarizations. RD analysis was performed, and four targets were detected using all three antennas, as can be seen in Figure 11. The study also highlighted the challenges and complexities of long-baseline bistatic radar, including timing synchronization, phase coherence, and the need for an accurate target ephemeris. To the best of our knowledge, orbit determination results are not included in the study.

The same RTG performed monostatic and bistatic radar imaging experiments on real on-orbit tumbling rocket bodies by utilizing the same network of transmitters and receivers [64–67]. These studies demonstrated advanced bistatic Doppler characterization across diverse imaging geometries and highlighted the successful implementation of specialized Doppler processing techniques, such as the Doppler superpulse algorithm. The study in [64] also outlines the methodology for accurately determining the rotation periods and maximum observed length extent of tumbling objects in space. Figure 12 displays the bistatic Doppler tomography reconstructions for the Atlas 5 Centaur rocket body in the L-band with an 8 MHz bandwidth based on WSRT data collection. The PP and OP channels are shown in Figure 12a,b, respectively. Table 4 compares the rotation period estimates for Delta-4 and Atlas 5, derived from multiple receivers and different processing methods. Recent studies have presented a novel range–Doppler–time (RDT) tensor processing technique for GEO characterization with narrow-band radar, which provides detailed Doppler information [68].

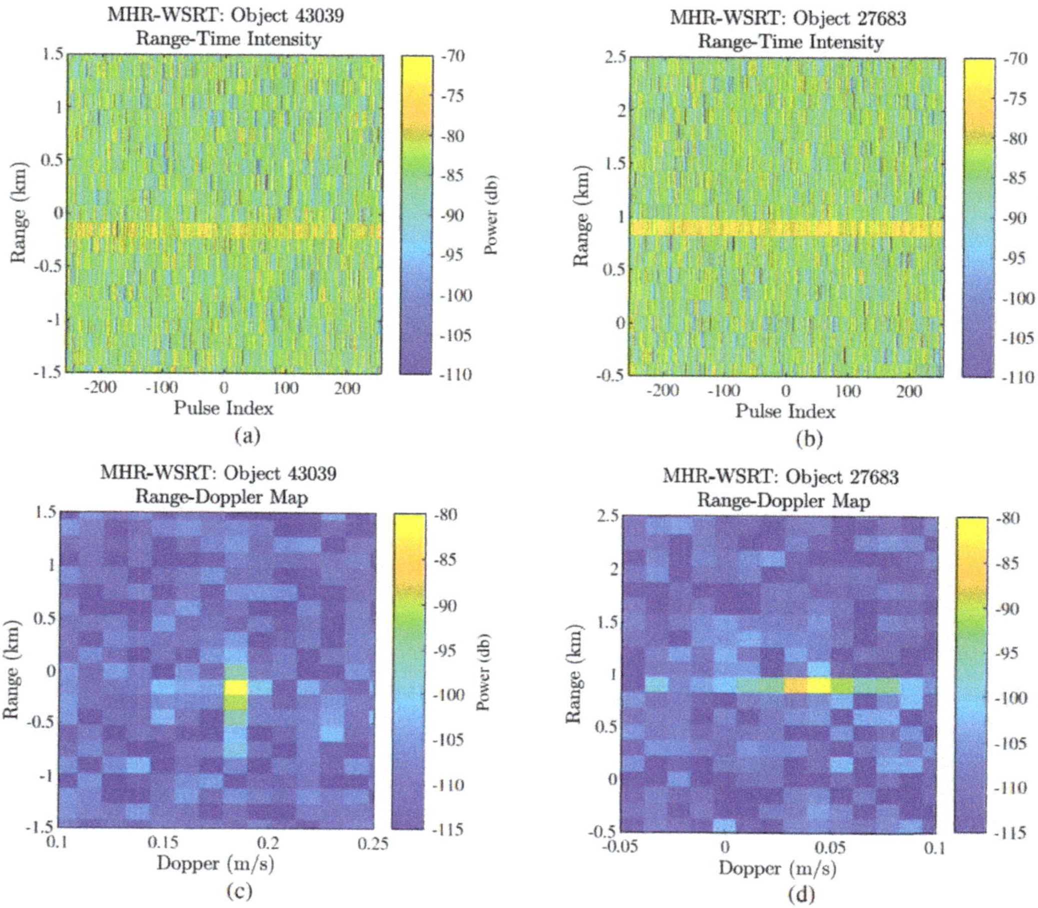

Figure 10. (**a**,**b**) Range—time intensity plots for objects 43039 and 27683, recorded during a single CPI in January 2020 using the MHR transmitter and WSRT receiver. These plots show signal strength variations over time and range. (**c**,**d**) Corresponding range–Doppler (RD) maps for the same targets. The origin (0,0) in each map is aligned with the expected range and Doppler shift of the respective object, aiding in accurate target identification and motion characterization [30].

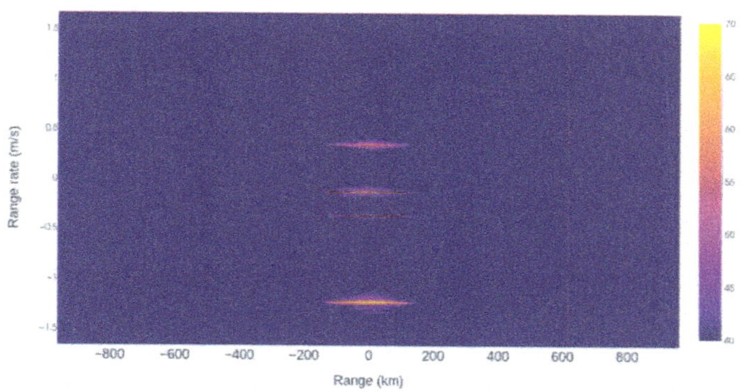

Figure 11. Range—Doppler map generated from the TIRA–Knockin bistatic radar data using a CPI of 512 pulses. The map clearly reveals four distinct targets detected during the January 2020 experimental campaign [62].

In another study, Riofrío et al. [69] discussed the performance analysis of ground-based long baseline radar for SSA by examining targets at different altitudes. The study examines the performance of a multistatic system considering one transmitter in three scenarios: a cluster of receivers, receivers spread throughout the world, and a combination of both. The authors used the multiple-input–multiple-output (MIMO) ambiguity function to analyze the system's performance, and simulations were performed for targets in LEO and GEO to check the feasibility of the proposed radar systems.

Table 4. The table shows the results of rotation period estimation using different methods: **TAC**—time–domain auto-correlation, **TFAC**—time–frequency domain auto-correlation, **STC**—slow-time cepstrum, **SFC**—slow frequency spectogram, **SIRTA**—spectrogram–inverse Radon transform-based algorithm [64]. Reprinted from IET Radar, Sonar, and Navigation, Copyright 2023, with permission from John Wiley and Sons.

Target	Receiver	TAC	TFAC	STC	SFS	SIRTA	Average
Delta-4	WSRT	164.2373	164.1280	164.2228	164.1748	164.4878	164.2501
	JBO-KN	164.0500	164.0257	164.0103	161.3476	164.6694	164.1889
	JBO-DE	164.2581	164.1588	164.2345	162.1518	164.3632	164.2537
	JBO-DA	164.2790	164.2472	164.3130	162.5722	164.9472	164.4466
	JBO-CM	164.1957	164.2717	164.3948	166.5494	164.5672	164.3573
Atlas 5	WSRT	33.6552	33.6567	33.6627	33.6651	33.6490	33.6577
	TIRA	33.6388	33.6507	33.6654	33.6221	34.8242	33.6442

JBO—Jodrell Bank observatory, KN—Knockin, DE—Defford, DA—Darnhall, CM—Cambridge.

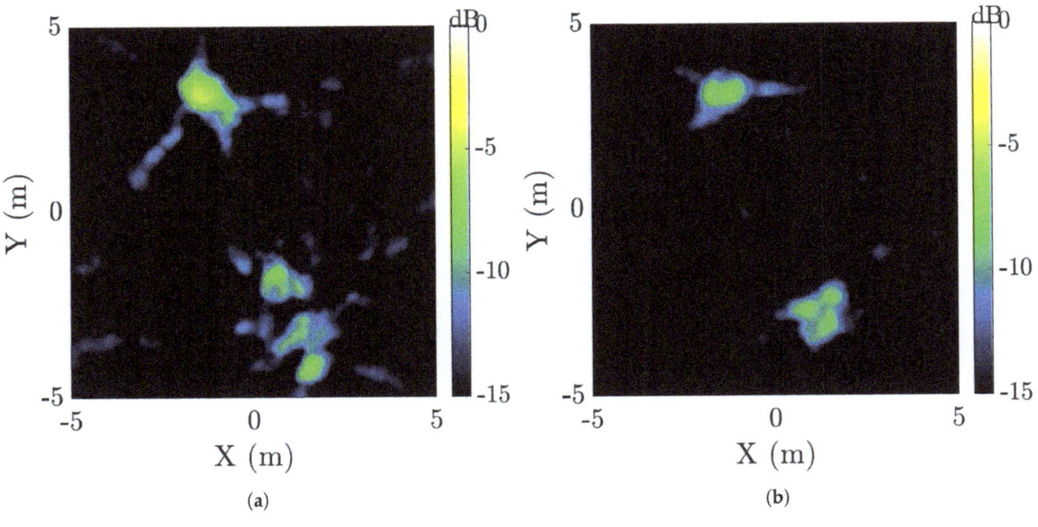

Figure 12. Bistatic Doppler tomography reconstructions of the Atlas 5 Centaur rocket body using L-band data with 8 MHz bandwidth, collected by WSRT. The results illustrate the target structure based on Doppler information: (**a**) PP—prime polarization channel, (**b**) OP—orthogonal polarization channel [64]. Reprinted from IET Radar, Sonar, and Navigation, Copyright 2023, with permission from John Wiley and Sons.

3.8. MeerKAT

The Meer Karoo Array Telescope (MeerKAT), located in the Northern Cape of South Africa, is part of the SKA network. It comprises 64 dishes intended for radio astronomy and serves as proof of technological advancement toward constructing a Square Kilometre Array (SKA), as shown in Figure 13. The MeerKAT radar is designed to operate at 1350 MHz

(L-Band) frequency with a bandwidth of 10 MHz. The Radar Remote Sensing Group (RRSG) at the University of Cape Town, South Africa, laid the essential groundwork with simulations for the Mission Planning Tool (MPT), which will be used at the MeerKAT radar to perform sensor scheduling and orbit determination analysis [70,71]. Agaba's PhD thesis [71] primarily concentrates on designing the comprehensive system operation and the radar signal processing for the proposed MeerKAT radar with the RRSG's Flexible Extensible Radar Simulator (FERS) [72]. In addition, different radar configurations, such as monostatic, quasi-monostatic, and single-input–multiple-output (SIMO), were analyzed to determine the best performance with MeerKAT. The results indicate that the multistatic radar configuration is the most appropriate and computationally efficient option compared with the bistatic and SIMO configurations. The research presented in this PhD thesis also focused on Inverse Synthetic Aperture Radar (ISAR) imaging, comparing simulated results with theoretical models while developing methods to correct range and Doppler ambiguities, thereby improving imaging precision [71].

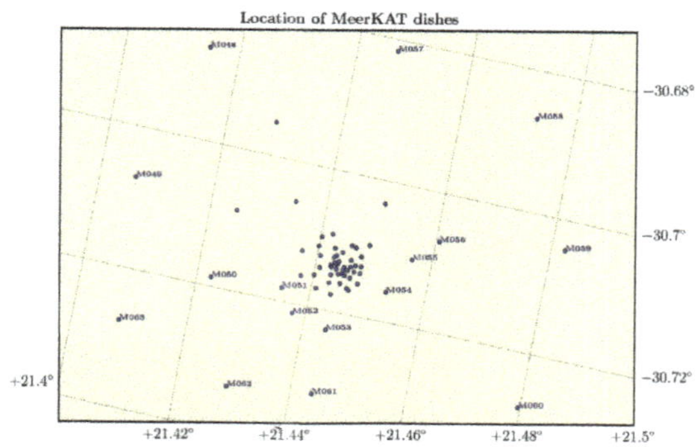

Figure 13. Map showing the layout of the MeerKAT radio telescope array. The receptors, labeled M000 to M063, are distributed across the site based on a two-dimensional Gaussian configuration [70].

Moreover, in another study, a bistatic system was proposed with a transmitter located at Denel Dynamics' Overberg Test Range in Arniston, Bredasdorp (Western Cape, South Africa) and a receiver at MeerKAT, separated by a baseline of approximately 450 km. A preliminary case study was conducted to demonstrate the operation of a bistatic passive radar system to observe the International Space Station [73]. The authors compared the capabilities of the MeerKAT radar to the BIRALES system, as described in Section 3.4, to estimate azimuth and elevation angles [70]. This comparison was necessary because the current design of MeerKAT does not include an angle of arrival estimation scheme. Nonetheless, the simulation results demonstrate the potential effectiveness of the MeerKAT radar in tracking LEO space debris. However, this must be validated through real-world measurement campaigns.

4. Radio Telescope Measurements with Illuminators of Opportunity

Typically, terrestrial broadcasting networks, including television and radio, comprise a distributed array of multiple transmitters. This configuration enables the simultaneous utilization of several transmitters as radar illumination sources above the receiver. Transmitting antennas are typically oriented towards the ground to maximize the power directed at the intended terrestrial targets. However, this design choice can present challenges for illuminating satellites, as considerable effort has been made to minimize outward radiation, with the main elevation sidelobes exhibiting power levels that are substantially lower, up to 15 dB, compared with the main beam [74]. Currently, there are at least two prominent facilities, the Murchison Widefield Array (MWA) located in Australia and part of the LOFAR telescope in Poland, which are actively working on developing and utilizing passive radar systems for space debris monitoring. These passive radar configurations leverage existing sources of illumination, such as FM and DVB signals, to detect and track space debris, thereby providing a cost-effective and expandable solution for this growing challenge. The MWA, like the MeerKAT radar, is part of the SKA project, and a detailed discussion is provided in Section 4.1. Measurement results from the LOFAR telescope are presented and discussed in detail in Section 4.2.

4.1. Murchison Widefield Array

The Murchison Widefield Array (MWA), a member of the trio of SKA Precursor telescopes, is situated at the Murchison Radio-astronomy Observatory in the Murchison Shire, located in the mid-west of Western Australia. This site was specifically chosen because of its exceptionally low radio-frequency interference. Operating in the low radio frequency range of 80–300 MHz, the MWA features a processed bandwidth of 30.72 MHz per linear polarization. It comprises 128 aperture arrays or tiles spread across an area with a diameter of approximately 3 km, as illustrated in Figure 14 [17,75]. The primary objectives of MWA include detecting neutral hydrogen from the epoch of re-ionization, investigating Earth's ionosphere, and creating maps of both galactic and extragalactic radio sources. Earlier demonstrations used MWA to detect ISS and the moon using reflected FM radio emissions, but observations were limited to angular domain (i.e., right ascension and declination) only [76,77]. The experimental campaign conducted in April 2015 used both the MWA and a Defense Science and Technology group line-of-sight receiver in Adelaide. The primary target was the ISS, which was illuminated by nearby FM radio transmissions, as described in the scenario outlined in [78]. Researchers in [78,79] developed a passive radar signal processing technique to derive the bistatic range and Doppler shift with a coherent processing interval of 1 s. This technique improved the detection performance by 25–40 dB compared with the previous case where only angle observations were possible [80]. Additionally, azimuth and elevation measurements were derived through interferometric processing. These four measurements were combined to perform IOD and tracking by using the Markov chain Monte Carlo algorithm.

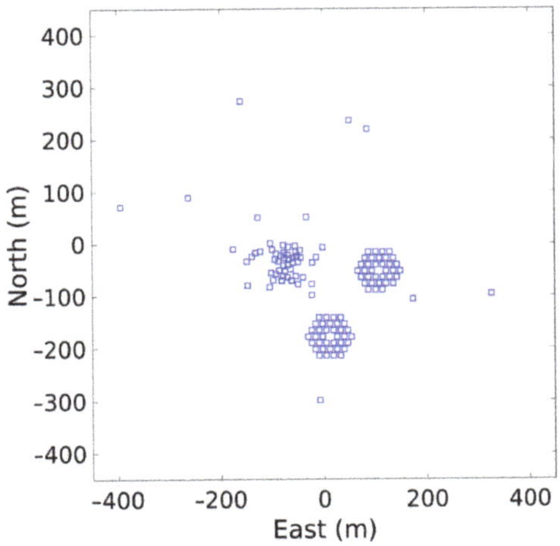

Figure 14. Layout of the Murchison Widefield Array (MWA). Each receiver is represented by a blue marker. The array consists of multiple tiles spread across the site to form a large interferometric baseline. Reproduced from [17].

Hennessy et al. [81] introduced the concept of Orbit Determination Before Detect (ODBD) for MWA, which constrains the search space to realistic orbital trajectories and improves detection efficiency. This algorithm incorporates orbital parameters into the radar ambiguity function and applies matched filtering to single-beam radar systems. The authors further validated the algorithm by applying it to a real measurement campaign at MWA conducted in 2018 and comparing it with the standard IOD method, such as Herrick-Gibbs, demonstrating the practical implementation of the algorithm, which further enhances the accuracy of debris detection [82]. While the authors proposed ways to constrain the search space, it may still require significant processing power for practical implementation.

Another measurement campaign was conducted in December 2019 using four transmitters (Geraldton, Perth, Albany, Mount Gambier) [17], and one of the detected targets, COSMOS 1707, was tracked for almost 90 s. The tracking results from the multistatic measurements are shown in Figure 15, with the top row as the covariance of position and velocity, and the bottom row shows the error estimates of position and velocity in comparison with the TLE ephemeris. The authors also mentioned that these results are typical for most of the objects detected at the MWA in a bistatic configuration. The measurements were combined at the orbit determination stage, and it can be seen in Figure 15 that the joint result from the Perth and Albany transmitters shows significant improvement in both position and velocity covariance. The mean error values, especially in the velocity, were reduced, which shows that the orbit accuracy improved with the combined data. In the same campaign, the tracking results of other targets, OPS 5721 (9415) and NADEZHDA (25567), are also published [17].

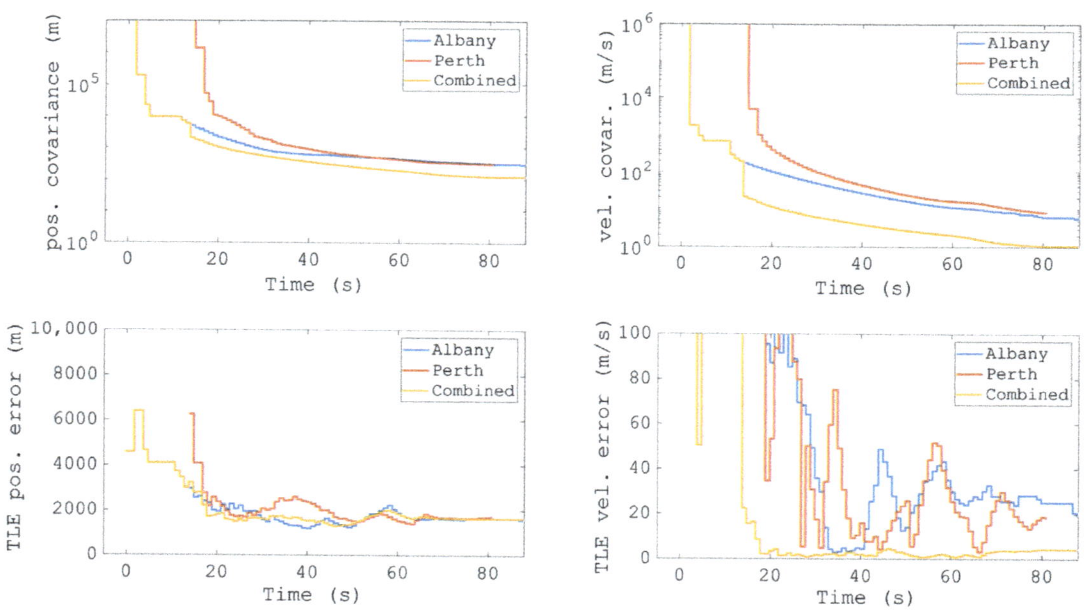

Figure 15. The resulting orbit predictions from the multistatic measurements of COSMOS 1707, conducted in December 2019 with the MWA radio telescope. The top rows show the covariance of the position estimate and the velocity estimate, and the bottom row shows the mean errors when compared with the Two-Line Element (TLE). Reproduced from [17].

Moreover, a rigorous assessment was conducted between 2019 and 2023 by investigators to extract precise angular measurements of satellites from spatially blurred detections produced by non-coherent passive radar techniques using the MWA [83,84]. By developing specialized signal processing methods, researchers were able to derive time-stamped angular coordinates for 32 observed satellite passes, which were then used to perform high-accuracy orbit determination through a least-squares fitting algorithm. The resulting median uncertainties of 860 m in the cross-track and 780 m in the in-track directions have been reported for LEO objects, which are comparable to the typical 1000 m uncertainty in publicly available TLE data [85,86]. Over the past decade, the development of MWA has been successful, with significant improvements in passive radar signal processing techniques that have enhanced the ability to refine the orbit determination of space objects. Advancements in MWA in this area have contributed to more accurate tracking and monitoring of the growing population of space debris, which poses escalating risks to operational satellites and future space missions.

4.2. LOFAR

The Netherlands Institute for Radio Astronomy, ASTRON, developed and built the LOw-Frequency ARray for radio astronomy (LOFAR), an international network of radio telescopes. LOFAR began operations in 2012 and currently comprises 52 independent stations. Of these, ASTRON manages 38 stations, whereas the remaining 14 are international stations situated across Europe [87,88]. Earlier research conducted by a group of researchers at the Warsaw University of Technology (Poland) explored the application of LOFAR radio telescopes as passive radar receivers [89]. Initially, the focus was on the possibility of detecting airplanes, but the research was expanded to include the observation of space objects traveling in LEO. The research team carried out experimental trials to validate the concept, utilizing a LOFAR station as the receiver and commercial digital audio broadcasting (DAB) transmitters as illuminators of opportunity for aerial object detection [90,91].

In a previous study, researchers employed the LOFAR station PL610 in Borówiec, Poland, and a proximate DVB-T transmitter to detect ISS [92,93]. The utilization of DVB-T offers the advantage of providing a broader bandwidth and potentially more precise range measurements compared with previously utilized DAB+ transmitters [93]. The cross-ambiguity function (CAF) calculated in [92] is displayed in Figure 16, showing its appearance both prior to and following the implementation of Direct Path Interference (DPI) and clutter cancellation techniques. Figure 16a clearly shows the presence of DPI and clutter elements. In contrast, Figure 16b illustrates the successful removal of these interferences, demonstrating the efficacy of the cancellation method. This enhancement results in a clearer CAF, which allows for a more precise identification of target signals in the vicinity of the zero bistatic range and velocity. This enabled the detection of weak signals from distant objects. The successful detection of ISS across a wide range highlights the capability of passive radar systems for space surveillance [94,95]. Recent studies have further demonstrated the potential of this approach by detecting Starlink satellites [96], which have significantly lower radar cross-sections than the ISS, showcasing the feasibility of exploiting the existing radio astronomy infrastructure to enhance space situational awareness.

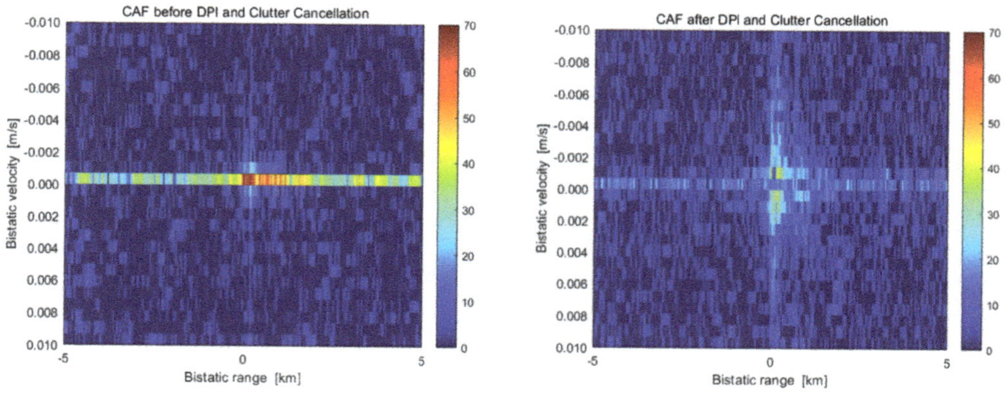

(a) (b)

Figure 16. (**a**) A zoom on cross -ambiguity function (CAF) near-zero bistatic range and velocity before direct-path interference (DPI) and clutter cancellation. (**b**) A zoom on cross-ambiguity function (CAF) near zero bistatic range and velocity after direct-path interference (DPI) and clutter cancellation, showing improved target signal clarity in the bistatic domain. The data were obtained using the LOFAR PL610 station with a DVB-T transmitter as an illuminator of opportunity for ISS detection [92]. Reprinted from IET Radar, Sonar, and Navigation, Copyright 2023, with permission from John Wiley and Sons.

5. Conclusions and Future Directions

An extensive review of past space debris monitoring campaigns using radio telescopes is conducted in this study. The challenges of monitoring space debris are compounded by the need to track fragments as small as 5 mm, as even at this size, they can cause significant damage to active satellites. Globally, more than 50 radar systems are currently dedicated to monitoring space objects and debris, and efforts are underway to expand this network of radar systems designed for space surveillance. Among these, bistatic radars with radio telescopes provide a balanced approach that combines high sensitivity and cost-effectiveness for space debris monitoring. In this study, an important perspective is explored by organizing the debris monitoring campaigns into two segments based on the type of transmitter used. A comparative summary of the radio telescopes discussed in this study is available in Table 5, listing relevant attributes including location, setup, transmission features, and detection performance. The analysis presented in Table 5 and the review of existing literature reveal that more campaigns have been conducted with active transmitter configurations, likely due to the advantage of higher transmitted power. Nevertheless, opportunistic transmitters are also favored because of their all-time availability and long exposure durations. This study included measurement campaigns for both LEO and GEO orbital regions, but most of the campaigns were conducted to monitor the LEO orbit, as it is highly populated with space debris. However, monitoring GEO targets is particularly challenging because of their large distance from the Earth, and it involves coordinating multiple countries to establish the necessary bistatic configuration and requires high transmission power, as emphasized in Section 3.7. Some of the radio telescopes discussed also operate in a multistatic configuration, which, despite introducing additional system complexity, significantly enhances overall sensitivity. This improvement, along with the transition from bistatic to multistatic configurations, provides advantages in terms of precise localization and measurement ambiguity reduction, as highlighted in Sections 3.6, 3.7 and 4.1 (indicated by † in Table 5).

There are a few limitations associated with conventional radio telescopes for space debris monitoring, and they have been highlighted in this manuscript. A common issue encountered when using radio telescopes is obtaining precise angular measurements, as most radio telescopes provide only range and Doppler measurements. Determining the orbit of an uncataloged object using only these two types of measurement can be challenging. However, some of the radio telescopes discussed here have addressed this problem through unique designs such as multibeam antennas and advanced data processing algorithms, as detailed in Sections 3.4 and 4.1, respectively. Single-beam radio telescopes still face difficulties in deriving accurate angular measurements, which could be an area of future research and improvement.

Table 5. Overview of the radio telescopes analyzed in this study, providing a comparative summary of the systems examined.

Configuration	Sensors Involved (Tx/Rx)	Country	Configuration	Operating Frequency	Waveform Type	Detection Performance	References
TIRA with Effelsberg Radio Telescope	TIRA (Tx), Effelsberg (Rx)	Germany	Active	1.33 GHz	Pulsed	Targets ≥ 1 cm RCS detectable at 1000 km	[34]
Goldstone orbital debris radar	GSSR (Tx), DSS (Rx)	USA	Active	8.56 GHz	Pulsed	Targets ≥ 3 mm RCS detectable at 1000 km	[22]
Arecibo Observatory	Arecibo Telescope (Rx)	USA	Active	430 MHz	Pulsed	Targets ≥ 5 mm RCS detectable at 1000 km	[23]
BIRALES	TRF (Tx), Medicina (Rx)	Italy	Active	410–415 MHz	CW	Targets ≥ 1 m RCS detectable at 1000 Km	[51]
BIRALET	TRF (Tx), SRT (Rx)	Italy	Active	410–415 MHz	CW	Targets ≥ −8 dBsm RCS detectable at 1000 km	[26]
Italian Multibistatic Radar†	Medicina (Rx), NOTO (Rx), SRT (Rx)	Italy	Active	-	-	N.A.	[27]
Long baseline bistatic radar†	TIRA (Tx), MHR (Tx), WSRT (Rx), E-Merlin (Rx), SRT (Rx)	Germany, USA, Netherlands, UK, Italy	Active	1.333 GHz (TIRA), 1.295 GHz (MHR)	Pulsed	Detection at GEO range	[62]
MeerKAT†	Denel Dynamics (Tx), MeerKAT (Rx)	South Africa	Interferometer-Active	1350 MHz	Pulsed	N.A.	[70,73]
Murchison Widefield Array (MWA) †	MWA (Rx)	Australia	Interferometer—Passive	80–300 MHz	Passive (FM broadcast)	Targets ≥ 0.5 m^2 RCS detectable at 1000 Km	[17]
LOFAR	LOFAR PL160 (Rx)	Poland	Interferometer—Passive	110–240 MHz	Passive (FM broadcast)	Targets ≥ 0.1 m^2 RCS detectable	[91,92]

† Multistatic-capable.

Several studies [11,97–99] suggested that integrating space-based radar and radio telescopes with existing space situational awareness programs can significantly improve debris tracking and collision avoidance strategies. This approach can enhance the detection accuracy, expand coverage beyond the low Earth orbit, and enable real-time monitoring of hazardous space debris. On the other hand, some of the radio telescopes discussed in this paper, such as BIRALES, MeerKAT, MWA, and LOFAR, operate based on the Very Long Baseline Interferometry (VLBI) principle or similar interferometric techniques, which are known to enhance spatial resolution and measurement accuracy. In addition, future research can expand the characterization of space objects using multistatic interferometric ISAR imaging, as very limited research has been conducted in this domain. A notable application of high-resolution radar imaging, utilizing similar principles, is the IoSiS experiment conducted by the German Space Agency, DLR, for imaging satellites and space debris in LEO [100,101]. Moreover, the mathematical model proposed in the recent literature [102] for the 3D imaging of resident space objects based on the interferometric approach needs to be further validated using real-world multistatic receiver networks. Its independence from the sensor system geometry makes it a promising approach for improving space object tracking and identification. This direction is further supported by the PRIDE experiment discussed in Section 3.6, which applied interferometric techniques for deep-space tracking and may inspire similar methodologies for space debris imaging in high-altitude orbits.

The comprehensive analysis in this manuscript reveals that space situational awareness efforts are primarily led by the USA and European countries, while relatively fewer experiments have been conducted in other parts of the world. Research institutions and space agencies in the Asia-Pacific region, such as ISRO and JAXA, have the potential to adopt innovative approaches and further advance this technology. It should be emphasized that one of the key reasons for the limited analysis reported so far is the lack of publicly available data, as some space agencies classify such work for security and strategic reasons. Although the USA and Europe have been significant contributors, engaging more stakeholders from other parts of the world could further propel progress in this domain. In addition to government initiatives, a few private companies are also contributing to space situational awareness, such as LeoLabs from the USA and Silentium Defense from Australia [103,104]. While LeoLabs uses phase-array technology to precisely track targets in space, the Maverick radar from Silentium Defense utilizes passive FM signals to detect and determine the orbit of debris. Moreover, advances in computational power and the refinement of machine learning and deep learning techniques have driven increasing interest in applying these methods to improve orbit prediction accuracy and resident space object classification [105–108]. Ongoing advancements in radar and radio telescope systems, along with improvements in data analysis and machine learning methods, are crucial for advancing future space debris tracking. By addressing existing technical issues and promoting international collaboration, researchers can devise more effective and scalable strategies to mitigate the increasing threats posed by space debris accumulation.

Author Contributions: Bibliographic research and analysis: B.A.; draft preparation: B.A.; writing, review, and editing: B.A., L.G., A.K. and M.M; supervision: L.G. and M.M. All authors have read and agreed to the published version of the manuscript.

Funding: This research received no external funding.

Data Availability Statement: Not applicable.

Acknowledgments: This paper and related research have been conducted during and with the support of the Italian National inter-university PhD program in Space Science and Technology.

Conflicts of Interest: The authors declare no conflicts of interest.

Abbreviations

The following abbreviations are used in this manuscript:

ISS	International Space Station
LEO	Low Earth Orbit
SSA	Space Situational Awareness

SST	Space Surveillance and Tracking
SSN	Space Surveillance Network
TLE	Two Line Element
LIDAR	Light Detection and Ranging
TIRA	Tracking and Imaging Radar
MWA	Murchison Widefield Array
RCS	Radar Cross-Section
SKA	Square Kilometer Array
VLBI	Very Long Baseline Interferometry
BIRALES	Bistatic Radar for LEO Survey
BIRALET	Bistatic Radar for LEO Tracking
SRT	Sardinia Radio Telescope
STO	Science and Technology Organization
GEO	Geostationary Earth Orbit
GESTRA	German Experimental Space Surveillance and Tracking Radar
NGAT	Next Generation Arecibo Telescope
TRF	Radio Frequency Transmitter
IOD	Initial Orbit Determination
CW	Continuous Wave
UKF	Unscented Kalman Filter
ECEF	Earth Centred Earth Fixed
NEO	Near-Earth Objects
PRIDE	Planetary Radio Interferometry and Doppler Experiment
LBBR	Long Baseline Bistatic Radar
RTG	Research Task Group
MHR	Millstone Hill Radar
WSRT	Westerbork Synthesis Radio Telescope
E merlin	enhanced MultiElement Remotely Linked Interferometer Network
MIMO	Multi Input Multi Output
MeerKAT	Meer Karoo Array Telescope
RRSG	Radar Remote Sensing Group
MPT	Mission Planning Tool
FERS	Flexible Extensible Radar Simulator
SIMO	Single-Input–Multiple-Output
ISAR	Inverse Synthetic Aperture Radar
ODBD	Orbit Determination Before Detect
CAF	Cross-Ambiguity Function
DPI	Direct Path Interference

References

1. Klinkrad, H.; Beltrami, P.; Hauptmann, S.; Martin, C.; Sdunnus, H.; Stokes, H.; Walker, R.; Wilkinson, J. The ESA space debris mitigation handbook 2002. *Adv. Space Res.* **2004**, *34*, 1251–1259. [CrossRef]
2. Kessler, D.J.; Johnson, N.L.; Liou, J.; Matney, M. The kessler syndrome: Implications to future space operations. *Adv. Astronaut. Sci.* **2010**, *137*, 2010.
3. Gasparini, G.; Miranda, V. *Space Situational Awareness: An Overview*; Springer: Berlin/Heidelberg, Germany, 2010.
4. Hu, Y.; Li, K.; Liang, Y.; Chen, L. Review on strategies of space-based optical space situational awareness. *J. Syst. Eng. Electron.* **2021**, *32*, 1152–1166.
5. Vallado, D.A.; Griesbach, J.D. Simulating space surveillance networks. In Proceedings of the Paper AAS 11-580 presented at the AAS/AIAA Astrodynamics Specialist Conference, Girdwood, Alaska, 31 July–4 August 2011.
6. Weeden, B.; Cefola, P.; Sankaran, J. Global space situational awareness sensors. In Proceedings of the AMOS Conference, Maui, HI, USA, 14–17 September 2010.
7. Vallado, D.; Crawford, P. SGP4 orbit determination. In Proceedings of the AIAA/AAS Astrodynamics Specialist Conference and Exhibit, Honolulu, Hawai, USA, 18–21 August 2008; p. 6770.
8. Mehrholz, D. Radar observations in low earth orbit. *Adv. Space Res.* **1997**, *19*, 203–212. [CrossRef]

9. Wilden, H.; Bekhti, N.B.; Hoffmann, R.; Kirchner, C.; Kohlleppel, R.; Reising, C.; Brenner, A.; Eversberg, T. GESTRA-recent progress, mode design and signal processing. In Proceedings of the 2019 IEEE International Symposium on Phased Array System & Technology (PAST), Waltham, MA, USA, 15–18 October 2019 ; pp. 1–8.
10. Bennet, F.; Rigaut, F.; Ritchie, I.; Smith, C. *Adaptive Optics to Enhance Tracking of Space Debris*; SPIE: Canberra, Australia, 2014; Volume 9148, p. 91481F.
11. Sajjad, N.; Mirshams, M.; Hein, A.M. Spaceborne and ground-based sensor collaboration: Advancing resident space objects' orbit determination for space sustainability. *Astrodynamics* **2024**, 8, 325–347. [CrossRef]
12. Wang, Y.; Wang, Q. The application of lidar in detecting space debris. In Proceedings of the 2008 International Conference on Optical Instruments and Technology: Optoelectronic Measurement Technology and Applications, Beijing, China, 16–19 November 2009; Volume 7160, pp. 464–470.
13. Muntoni, G.; Schirru, L.; Pisanu, T.; Montisci, G.; Valente, G.; Gaudiomonte, F.; Serra, G.; Urru, E.; Ortu, P.; Fanti, A. Space debris detection in low Earth orbit with the Sardinia radio telescope. *Electronics* **2017**, *6*, 59. [CrossRef]
14. Kraus, J.D.; Tiuri, M.; Räisänen, A.V.; Carr, T.D. *Radio Astronomy*; McGraw-Hill: New York, NY, USA, 1966; Volume 66.
15. Muntoni, G.; Montisci, G.; Pisanu, T.; Andronico, P.; Valente, G. Crowded space: A review on radar measurements for space debris monitoring and tracking. *Appl. Sci.* **2021**, *11*, 1364. [CrossRef]
16. Letsch, R.; Leushacke, L.; Rosebrock, J.; Jehn, R.; Krag, H.; Keller, R. First Results from the Multibeam Bistatic Beampark Experiment at FGAN. In Proceedings of the 5th European Conference on Space Debris, Darmstadt, Germany, 30 March–2 April 2009; Volume 33.
17. Hennessy, B.; Rutten, M.; Young, R.; Tingay, S.; Summers, A.; Gustainis, D.; Crosse, B.; Sokolowski, M. Establishing the capabilities of the Murchison Widefield Array as a passive radar for the surveillance of space. *Remote Sens.* **2022**, *14*, 2571. [CrossRef]
18. Skolnik, M.I. Introduction to radar. *Radar Handb.* **1962**, *2*, 21.
19. Markiewicz, T.G.; Wesołowski, K.W. Cryogenic Cooling in Wireless Communications. *Entropy* **2019**, *21*, 832. [CrossRef]
20. Willis, N.J. *Bistatic Radar*; SciTech Publishing: Raleigh, NC, USA, 2005; Volume 2.
21. Henry, J.K.; Narayanan, R.M.; Singla, P. Radar Cross-Section Modeling of Space Debris. In Proceedings of the International Conference on Dynamic Data Driven Applications Systems, Cambridge, MA, USA, 6–10 October 2022 ; pp. 81–94.
22. Matney, M.; Goldstein, R.; Kessler, D.; Stansbery, E. Recent results from Goldstone orbital debris radar. *Adv. Space Res.* **1999**, *23*, 5–12. [CrossRef]
23. Thompson, T.W.; Goldstein, R.M.; Campbell, D.B.; Stansbery, E.G.; Potter, A.E. Radar detection of centimeter-sized orbital debris: Preliminary Arecibo observations at 12.5-CM wavelength. *Geophys. Res. Lett.* **1992**, *19*, 257–259. [CrossRef]
24. Pupillo, G.; Bartolini, M.; Cevolani, G.; Di Martino, M.; Falkovich, I.; Konovalenko, A.; Malevinskij, S.; Montebugnoli, S.; Nabatov, A.; Pluchino, S.; et al. Space debris observational test with the Medicina-Evpatoria bistatic radar. *Mem. Della Soc. Astron. Ital.* **2008**, *12*, 44.
25. Pisanu, T.; Schirru, L.; Urru, E.; Maxia, P. Status of the Sardinia Radio Telescope as a Receiver of the BIRALET bi-Static Radar for Space Debris Observations. In Proceedings of the Ground-Based and Airborne Telescopes VIII, SPIE, Online, 14–18 December 2020; Volume 11445, pp. 1310–1325.
26. Pisanu, T.; Muntoni, G.; Schirru, L.; Ortu, P.; Urru, E.; Montisci, G. Recent advances of the BIRALET system about space debris detection. *Aerospace* **2021**, *8*, 86. [CrossRef]
27. Cataldo, D.; Gentile, L.; Ghio, S.; Giusti, E.; Tomei, S.; Martorella, M. Multibistatic radar for space surveillance and tracking. *IEEE Aerosp. Electron. Syst. Mag.* **2020**, *35*, 14–30. [CrossRef]
28. Gentile, L.; Martorella, M. LEO RSO Initial Orbit Determination using Range-only Multi-Bistatic Radar measurements. In Proceedings of the 2018 SpaceOps Conference, Marseille, France, 28 May–1 June 2018 ; p. 2717.
29. Ghio, S.; Martorella, M.; Staglianò, D.; Petri, D.; Lischi, S.; Massini, R. Experimental comparison of radon domain approaches for resident space object's parameter estimation. *Sensors* **2021**, *21*, 1298. [CrossRef]
30. Welch, S.; Hogan, G.; Morrison, R.; Bassa, C.; Pisanu, T. Long baseline radar bistatic measurements of geostationary satellites. In Proceedings of the 2022 19th European Radar Conference (EuRAD), Milan, Italy, 28–30 September 2022 ; pp. 1–4.
31. Li, J.; Gao, P.; Shen, M.; Zhao, Y. The capability analysis of the bistatic radar system based on Tianlai radio array for space debris detection. *Adv. Space Res.* **2019**, *64*, 1652–1661. [CrossRef]
32. Henault, S.; Guimond, J.F. Orbit determination of a LEO satellite with passive RF observation of a single pass by two collocated antennas. *Adv. Space Res.* **2025**, *75*, 5014–5025. [CrossRef]
33. Klare, J.; Behner, F.; Carloni, C.; Cerutti-Maori, D.; Fuhrmann, L.; Hoppenau, C.; Karamanavis, V.; Laubach, M.; Marek, A.; Perkuhn, R.; et al. The Future of Radar Space Observation in Europe—Major Upgrade of the Tracking and Imaging Radar (TIRA). *Remote Sens.* **2024**, *16*, 4197. [CrossRef]
34. Mehrholz, D.; Leushacke, L.; Jehn, R. The COBEAM-1/96 experiment. *Adv. Space Res.* **1999**, *23*, 23–32. [CrossRef]

35. Ruiz, G.; Leushacke, L.; Jehn, R.; Keller, R. Improved FGAN/MPIFR bi-static debris observation campaign: Experiment outline, analysis algorithms and first results. In Proceedings of the 57th International Astronautical Congress, Valencia, Spain, 2–6 October 2006 ; p. B6–1.
36. Flegel, S.; Letsch, K.; Krag, H. Theoretical analysis of south-staring Beampark configurations for the TIRA system. *CEAS Space J.* **2015**, *7*, 375–387. [CrossRef]
37. Ruiz, G.; Leushacke, L.; Rosebrock, J. Algorithms for multi-beam receiver data analysis. In Proceedings of the 4th European Conference on Space Debris, Darmstadt, Germany, 18–20 April 2005; Volume 587, p. 89.
38. Letsch, K.; Leushacke, L.; Rosebrock, J.; Krag, H.; Keller, R. Validation of revised multibeam estimation algorithms with data from TIRA/Effelsberg bistatic beampark experiments. In Proceedings of the 6th European Conference on Space Debris, Darmstadt, Germany, 22–25 April 2013; Volume 723, p. 126.
39. Manis, A.; Arnold, J.A.; Murray, J.; Buckalew, B.; Cruz, C.; Matney, M. An Overview of Ground-based Radar and Optical Measurements Utilized by the NASA Orbital Debris Program Office. In Proceedings of the 2nd International Orbital Debris Conference (IOC II), Houston, TX, USA, 4–7 December 2023.
40. Stokely, C.; Stansbery, E.; Goldstein, R. Debris flux comparisons from the Goldstone Radar, Haystack Radar, and Hax Radar prior, during, and after the last solar maximum. *Adv. Space Res.* **2009**, *44*, 364–370. [CrossRef]
41. Murray, J.; Miller, R.; Matney, M.; Anz-Meador, P.; Kennedy, T. Recent results from the goldstone orbital debris radar: 2016–2017. In Proceedings of the International Orbital Debris (IOC) Conference, Sugar Land, TX, USA, 9–12 December 2019.
42. Murray, J.; Jenet, F. The arecibo observatory as an instrument for investigating orbital debris: Legacy and next generation performance. *Planet. Sci. J.* **2022**, *3*, 52. [CrossRef]
43. Pisanu, T.; Schirru, L.; Urru, E.; Gaudiomonte, F.; Ortu, P.; Bianchi, G.; Bortolotti, C.; Roma, M.; Muntoni, G.; Montisci, G.; et al. Upgrading the Italian BIRALES system to a pulse compression radar for space debris range measurements. In Proceedings of the 2018 22nd International Microwave and Radar Conference (MIKON), Poznan, Poland, 14–17 May 2018; pp. 317–320. [CrossRef]
44. Losacco, M.; Di Lizia, P.; Massari, M.; Naldi, G.; Pupillo, G.; Bianchi, G.; Siminski, J. Initial orbit determination with the multibeam radar sensor BIRALES. *Acta Astronaut.* **2020**, *167*, 374–390. [CrossRef]
45. Bianchi, G.; Montaruli, M.F.; Roma, M.; Mariotti, S.; Di Lizia, P.; Maccaferri, A.; Facchini, L.; Bortolotti, C.; Minghetti, R. A new concept of transmitting antenna on bi-static radar for space debris monitoring. In Proceedings of the 2022 International Conference on Electrical, Computer, Communications and Mechatronics Engineering (ICECCME), Maldives, Maldives, 16–18 November 2022. . [CrossRef]
46. Di Lizia, P.; Massari, M.; Losacco, M.; Bianchi, G.; Mattana, A.; Pupillo, G.; Bortolotti, C.; Roma, M.; Morselli, A.; Armellin, R.; et al. Performance assessment of the multibeam radar sensor birales for space surveillance and tracking. In Proceedings of the 7th European Conference on Space Debris, Darmstadt, Germany, 18–21 April 2017.
47. Montaruli, M.F.; De Luca, M.A.; Massari, M.; Bianchi, G.; Magro, A. Operational Angular Track Reconstruction in Space Surveillance Radars through an Adaptive Beamforming Approach. *Aerospace* **2024**, *11*, 451. . [CrossRef]
48. Podda, A.; Casu, S.; Coppola, A.; Protopapa, F.; Sergiusti, A.L.; Pisanu, T.; Urru, E.; Schirru, L.; Ortu, P.; Gaudiomonte, F.; et al. Exploitation of bi-static radar architectures for LEO Space Debris surveying and tracking: The BIRALES/BIRALET project. In Proceedings of the 2020 IEEE Radar Conference (RadarConf20), Florence, Italy, 21–25 September 2020; pp. 1–6.
49. Facchini, L.; Montaruli, M.F.; Di Lizia, P.; Massari, M.; Pupillo, G.; Naldi, G.; Bianchi, G. Resident space object track reconstruction using a multireceiver radar system. In Proceedings of the 8th European Conference on Space Debris, ESA/ESOC, Darmstadt, Germany, 20–23 April 2021; pp. 1–8.
50. Montaruli, M.; Facchini, L.; Lizia, P.D.; Massari, M.; Pupillo, G.; Bianchi, G.; Naldi, G. Adaptive track estimation on a radar array system for space surveillance. *Acta Astronaut.* **2022**, *198*, 111–123. [CrossRef]
51. Losacco, M.; Di Lizia, P.; Mauro, M.; Bianchi, G.; Pupillo, G.; Mattana, A.; Naldi, G.; Bortolotti, C.; Roma, M.; Schiaffino, M.; et al. The multibeam radar sensor BIRALES: Performance assessment for space surveillance and tracking. In Proceedings of the 2019 IEEE Aerospace Conference, Big Sky, MT, USA, 2–9 March 2019.
52. Muciaccia, A.; Facchini, L.; Montaruli, M.F.; Purpura, G.; Detomaso, R.; Colombo, C.; Massari, M.; Di Lizia, P.; Di Cecco, A.; Salotti, L.; et al. Observation and analysis of Cosmos 1408 fragmentation. In Proceedings of the 73rd International Astronautical Congress (IAC 2022), Paris, France, 18–22 September 2022.
53. Montaruli, M.F.; Facchini, L.; Faraco, N.; Di Lizia, P.; Massari, M.; Bianchi, G.; Bortolotti, C.; Maccaferri, A.; Roma, M.; Peroni, M.; et al. Third Long-March 5B re-entry campaign through Italian space surveillance radars. *Adv. Space Res.* **2025**, *75*, 2139–2155. [CrossRef]
54. Wan, E.A.; Van Der Merwe, R. The unscented Kalman filter for nonlinear estimation. In Proceedings of the Proceedings of the IEEE 2000 Adaptive Systems for Signal Processing, Communications, and Control Symposium (Cat. No. 00EX373), Lake Louise, AB, Canada, 4 October 2000; pp. 153–158.

55. Cutajar, D.; Magro, A.; Borg, J.; Adami, K.; Bianchi, G.; Pupillo, G.; Mattana, A.; Naldi, G.; Bortolotti, C.; Perini, F.; et al. PyBIRALES: A Radar Data Processing Backend for the Real-Time Detection of Space Debris. *J. Astron. Instrum.* **2020**, *9*, 2050003. . [CrossRef]
56. De Luca, M.; Demuru, R.; Sangaletti, G.; Mesiano, L.; Boreanaz, I. The new transmitting antenna for BIRALES. In *Aeronautics and Astronautics: AIDAA XXVII International Congress*; Materials Research Forum LLC: Millersville, PA, USA 2023; Volume 37, p. 474.
57. Schirru, L.; Muntoni, G.; Pisanu, T.; Urru, E.; Valente, G.; Gaudiomonte, F.; Ortu, P.; Melis, A.; Concu, R.; Bianchi, G.; et al. Upgrading the Sardinia Radio Telescope to a Bistatic Tracking Radar for Space Debris. In Proceedings of the 1st IAA Space Situational Awareness Conference, Orlando, FL, USA, 13–15 November 2017.
58. Schirru, L.; Pisanu, T.; Navarrini, A.; Urru, E.; Gaudiomonte, F.; Ortu, P.; Montisci, G. Advantages of using a C-band Phased Array Feed as a receiver in the Sardinia Radio Telescope for Space Debris monitoring. In Proceedings of the 2019 IEEE 2nd Ukraine Conference on Electrical and Computer Engineering (UKRCON), Lviv, Ukraine, 2–6 July 2019; pp. 133–136.
59. Ahuja, B.; Gentile, L.; Marotrella, M. Improving Satellite Position And Velocity Calculation During Low Thrust Maneuvers Using Multi-Bistatic Radar And Unscented Kalman Filter. In Proceedings of the 4th IAA Space Situational Awareness Conference, Daytona Beach, FL, USA, 8–10 May 2024.
60. Pupillo, G.; Righini, S.; Orosei, R.; Bortolotti, C.; Maccaferri, G.; Roma, M.; Mastrogiuseppe, M.; Pisanu, T.; Schirru, L.; Cicalò, S.; et al. Toward a European Facility for Ground-Based Radar Observations of Near-Earth Objects. *Remote Sens.* **2023**, *16*, 38. [CrossRef]
61. Gurvits, L.I.; Cimò, G.; Dirkx, D.; Pallichadath, V.; Akins, A.; Altobelli, N.; Bocanegra-Bahamon, T.M.; Cazaux, S.M.; Charlot, P.; Duev, D.A.; et al. Planetary radio interferometry and Doppler experiment (PRIDE) of the JUICE mission. *Space Sci. Rev.* **2023**, *219*, 79. [CrossRef]
62. Welch, S.; Hogan, G.; Cerutti-Maori, D.; Garrington, S.; Morrison, R.; Otten, M.; Bassa, C.; Pallecaros, N.; van Dorp, P.; Harrison, P.; et al. Long Baseline Bistatic Radar for Space Situational Awareness. In Proceedings of the NATO SET-SCI-297 Specialists' Meeting on Space Sensors and Space Situational Awareness, Interlaken, Switzerland, 10–11 October 2022.
63. Früh, C.; Schildknecht, T. Accuracy of two-line-element data for geostationary and high-eccentricity orbits. *J. Guid. Control. Dyn.* **2012**, *35*, 1483–1491. [CrossRef]
64. Serrano, A.; Kobsa, A.; Uysal, F.; Cerutti-Maori, D.; Ghio, S.; Kintz, A.; Morrison, R.L., Jr.; Welch, S.; van Dorp, P.; Hogan, G.; et al. Long baseline bistatic radar imaging of tumbling space objects for enhancing space domain awareness. *IET Radar Sonar Navig.* **2024**, *18*, 598–619. [CrossRef]
65. Bosma, D.A.; Uysal, F. First Results of Attitude Estimation Using RCS Time Series for Space Domain Awareness. In Proceedings of the 2024 IEEE Radar Conference (RadarConf24), Denver, CO, USA, 6–10 May 2024; pp. 1–6.
66. Uysal, F.; Van Dorp, P.; Serrano, A.; Kobsa, A.; Ghio, S.; Kintz, A.; Bassa, C.; Garrington, S.; Cuenca, M.C.; Otten, M.; et al. Large baseline bistatic radar imaging for space domain awareness. In Proceedings of the 2023 IEEE International Radar Conference (RADAR), Sydney, Australia, 6–10 November 2023; pp. 1–6.
67. Kintz, A.; Hogan, G.; Kobsa, A.; Garrington, S.; Serrano, A.; Welch, S.; Morrison, R.; Ghio, S.; Bassa, C.; Harrison, P.; et al. Bistatic Radar for Military Space Domain Awareness at Geosynchronous Orbits. In Proceedings of the NATO SET-311 10th Military Sensing Symposium, London, UK, 19–21 April 2023.
68. Serrano, A.; Capper, J.; Morrison, R.L., Jr.; Abouzahra, M.D. Range-Doppler-Time Tensor Processing for Deep-Space Satellite Characterization Using Narrowband Radar. *Remote Sens.* **2024**, *16*, 1374. [CrossRef]
69. Díaz Riofrío, S.; Da Graça Marto, S.; Ilioudis, C.; Vasile, M.; Clemente, C. Performance analysis of ground-based long baseline radar distributed systems for space situational awareness. *IET Radar Sonar Navig.* **2024**, *18*, 586–597. [CrossRef]
70. Dhondea, A.R. Mission Planning Tool for Space Debris Studies with the MeerKAT Radar. Ph.D. Thesis, University of Cape Town, Engineering and the Built Environment, Department of Electrical Engineering, Cape Town, South Africa, 2018.
71. Agaba, D. System Design of the MeerKAT L-Band 3D Radar for Monitoring Near Earth Objects. Ph.D. Thesis, University of Cape Town, Faculty of Engineering & the Built Environment, Department of Electrical Engineering, Cape Town, South Africa, 2017.
72. Brooker, M. *The Design and Implementation of a Simulator for Multistatic Radar Systems.* 2008. Available online: http://hdl.handle.net/11427/5253 (accessed on 29 April 2025).
73. Dhondea, A.; Inggs, M. Mission planning tool for space debris detection and tracking with the MeerKAT radar. In Proceedings of the 2019 IEEE Radar Conference (RadarConf), Boston, MA, USA, 22–26 April 2019; pp. 1–6.
74. O'Hagan, D.W.; Griffiths, H.D.; Ummenhofer, S.M.; Paine, S.T. Elevation pattern analysis of common passive bistatic radar illuminators of opportunity. *IEEE Transactions Aerosp. Electron. Syst.* **2017**, *53*, 3008–3019. [CrossRef]
75. Lonsdale, C.J.; Cappallo, R.J.; Morales, M.F.; Briggs, F.H.; Benkevitch, L.; Bowman, J.D.; Bunton, J.D.; Burns, S.; Corey, B.E.; DeSouza, L.; et al. The murchison widefield array: Design overview. *Proc. IEEE* **2009**, *97*, 1497–1506. [CrossRef]
76. Tingay, S.J.; Goeke, R.; Bowman, J.D.; Emrich, D.; Ord, S.M.; Mitchell, D.A.; Morales, M.F.; Booler, T.; Crosse, B.; Wayth, R.B.; et al. The Murchison widefield array: The square kilometre array precursor at low radio frequencies. *Publ. Astron. Soc. Aust.* **2013**, *30*, e007. [CrossRef]

77. McKinley, B.; Briggs, F.; Kaplan, D.L.; Greenhill, L.; Bernardi, G.; Bowman, J.D.; de Oliveira-Costa, A.; Tingay, S.J.; Gaensler, B.M.; Oberoi, D.; et al. Low-frequency observations of the Moon with the Murchison widefield array. *Astron. J.* **2012**, *145*, 23. [CrossRef]
78. Palmer, J.E.; Hennessy, B.; Rutten, M.; Merrett, D.; Tingay, S.; Kaplan, D.; Tremblay, S.; Ord, S.M.; Morgan, J.; Wayth, R.B. Surveillance of Space using passive radar and the Murchison Widefield Array. In Proceedings of the 2017 IEEE radar conference (Radarconf), Seattle, WA, USA, 8–12 May 2017; pp. 1715–1720.
79. Palmer, J.E.; Harms, H.A.; Searle, S.J.; Davis, L. DVB-T passive radar signal processing. *IEEE Trans. Signal Process.* **2012**, *61*, 2116–2126. [CrossRef]
80. Tingay, S.J.; Kaplan, D.L.; McKinley, B.; Briggs, F.; Wayth, R.B.; Hurley-Walker, N.; Kennewell, J.; Smith, C.; Zhang, K.; Arcus, W.; et al. On the detection and tracking of space debris using the Murchison Widefield Array. I. Simulations and test observations demonstrate feasibility. *Astron. J.* **2013**, *146*, 103. [CrossRef]
81. Hennessy, B.; Rutten, M.; Tingay, S.; Young, R. Orbit determination before detect: Orbital parameter matched filtering for uncued detection. In Proceedings of the 2020 IEEE International Radar Conference (RADAR), Washington, DC, USA, 28–30 April 2020; pp. 889–894.
82. Hennessy, B.; Tingay, S.; Young, R.; Rutten, M.; Crosse, B.; Hancock, P.; Sokolowski, M.; Johnston-Hollitt, M.; Kaplan, D.L. Uncued detection and initial orbit determination from short observations with the Murchison Widefield array. *IEEE Aerosp. Electron. Syst. Mag.* **2021**, *36*, 16–30. [CrossRef]
83. Prabu, S.; Hancock, P.; Zhang, X.; Tingay, S.J.; Hodgson, T.; Crosse, B.; Johnston-Hollitt, M. Improved sensitivity for space domain awareness observations with the murchison widefield array. *Adv. Space Res.* **2022**, *70*, 812–824. [CrossRef]
84. Prabu, S.; Hancock, P.; Zhang, X.; Tingay, S.J. A low-frequency blind survey of the low Earth orbit environment using non-coherent passive radar with the Murchison widefield array. *Publ. Astron. Soc. Aust.* **2020**, *37*, e052. [CrossRef]
85. Prabu, S.; Hancock, P.J.; Zhang, X.; Tingay, S.J. The development of non-coherent passive radar techniques for space situational awareness with the Murchison Widefield Array. *Publ. Astron. Soc. Aust.* **2020**, *37*, e010. [CrossRef]
86. Prabu, S.; Hancock, P.; Zhang, X.; Tingay, S. Demonstration of orbit determination for LEO objects using the Murchison Widefield Array. *Adv. Space Res.* **2023**, *72*, 3282–3296. [CrossRef]
87. de Vos, M.; Gunst, A.W.; Nijboer, R. The LOFAR telescope: System architecture and signal processing. *Proc. IEEE* **2009**, *97*, 1431–1437. [CrossRef]
88. Heald, G.; McKean, J.; Pizzo, R. *Low Frequency Radio Astronomy and the LOFAR Observatory*; Springer: Berlin/Heidelberg, Germany, 2018; Volume 10, pp. 978–983.
89. Kłos, J.; Droszcz, A.; Jędrzejewski, K.; Kulpa, K.; Pożoga, M. On the possibility of using LOFAR radio telescope for passive radiolocation. In Proceedings of the 2020 21st International Radar Symposium (IRS), Warsaw, Poland, 5–8 October 2020; pp. 73–76.
90. Malanowski, M.; Jędrzejewski, K.; Misiurewicz, J.; Kulpa, K.; Gromek, A.; Kłos, J.; Droszcz, A.; Pożoga, M. Passive radar based on LOFAR radio telescope for air and space target detection. In Proceedings of the 2021 IEEE Radar Conference (RadarConf21), Atlanta, GA, USA, 7–14 May 2021; pp. 1–6.
91. Kłos, J.; Jędrzejewski, K.; Droszcz, A.; Kulpa, K.; Pożoga, M.; Misiurewicz, J. Experimental Verification of the Concept of Using LOFAR Radio-Telescopes as Receivers in Passive Radiolocation Systems. *Sensors* **2021**, *21*, 2043. [CrossRef]
92. Jędrzejewski, K.; Malanowski, M.; Kulpa, K.; Pożoga, M. Experimental verification of passive radar space object detection with a single low-frequency array radio telescope. *IET Radar Sonar Navig.* **2024**, *18*, 68–77. [CrossRef]
93. Jędrzejewski, K.; Malanowski, M.; Kulpa, K.; Pożoga, M. First experimental trials of passive DVB-T based space object detection with a single LOFAR radio telescope. In Proceedings of the International Conference on Radar Systems (RADAR 2022), Edinburgh, UK, 24–27 October 2022.
94. Jędrzejewski, K.; Pożoga, M.; Modrzewski, A.; Karwacki, M.; Kulpa, K.; Rothkaehl, H. Bistatic Detection of LEO Satellites from Very Long Distances Using LOFAR Radio Telescope. In Proceedings of the 2024 International Radar Symposium (IRS), Wroclaw, Poland, 2–4 July 2024; pp. 402–407.
95. Jędrzejewski, K.; Kulpa, K.; Malanowski, M.; Pożoga, M.; Wójtowicz, P. Exploring the Feasibility of Detecting LEO Space Objects in Passive Radar Without Prior Orbit Parameter Information. In Proceedings of the 2024 IEEE Radar Conference (RadarConf24), Denver, CO, USA, 6–10 May 2024; pp. 1–6.
96. Jedrzejewski, K.; Pozoga, M.; Wojtowicz, P.; Malanowski, M.; Kulpa, K.; Rothkaehl, H. Passive Observation of Starlink Satellites Using LOFAR Radio Telescope. In Proceedings of the 2024 International Radar Symposium (IRS), Wroclaw, Poland, 2–4 July 2024; pp. 408–413.
97. De Maio, A.; Maffei, M.; Aubry, A.; Farina, A. Effects of plasma media with weak scintillation on the detection performance of spaceborne radars. *IEEE Trans. Geosci. Remote Sens.* **2021**, *60*, 1–13. [CrossRef]
98. Maffei, M.; Aubry, A.; De Maio, A.; Farina, A. Spaceborne radar sensor architecture for debris detection and tracking. *IEEE Trans. Geosci. Remote Sens.* **2020**, *59*, 6621–6636. [CrossRef]

99. Maffei, M.; Aubry, A.; De Maio, A.; Farina, A. Emerging Trends in Radar: Spaceborne Radar for Surveillance. *IEEE Aerosp. Electron. Syst. Mag.* **2025**, 1–4. [CrossRef]
100. Anger, S.; Jirousek, M.; Dill, S.; Peichl, M. IoSiS—A high performance experimental imaging radar for space surveillance. In Proceedings of the 2019 IEEE Radar Conference (RadarConf), Boston, MA, USA, 22–26 April 2019; pp. 1–4.
101. Hochberg, F.; Jirousek, M.; Anger, S.; Peichl, M. Expanding Imaging of Satellites in Space (IoSiS): A Feasibility Study on the 3-Dimensional Imaging of Satellites Using Interferometry and Tomography. *Electronics* **2024**, *13*, 4914. [CrossRef]
102. Thindlu Rudrappa, M.; Albrecht, M.; Knott, P. Characterisation of resident space objects using multistatic interferometric inverse synthetic aperture radar imaging. *IET Radar Sonar Navig.* **2024**, *18*, 620–634. [CrossRef]
103. Griffith, N.; Lu, E.; Nicolls, M.; Park, I.; Rosner, C. Commercial Space Tracking Services for Small Satellites. In Proceedings of the Small Satellite Conference, Utah State University, Logan, UT, USA, 5–8 August 2019.
104. Finch, D.; Palmer, J.; Cooper, J.; Oltrogge, D.; Johnson, T. Assessing passive radar for LEO SSA. In Proceedings of the Advanced Maui Optical and Space Surveillance Technologies Conference, Maui, HI, USA, 27–30 September 2022.
105. Dumitrescu, F.; Ceachi, B.; Truică, C.O.; Trăscău, M.; Florea, A.M. A novel deep learning-based relabeling architecture for space objects detection from partially annotated astronomical images. *Aerospace* **2022**, *9*, 520. [CrossRef]
106. Hou, P.; Wang, Z.; Jiang, Y.; Zhu, J.; Chu, Q.; Li, J. Research Advancements in Artificial Intelligence for Space Situational Awareness. In Proceedings of the 2023 13th International Conference on Information Science and Technology (ICIST), Cairo, Egypt, 8–14 December 2023; pp. 568–577.
107. Massimi, F.; Ferrara, P.; Petrucci, R.; Benedetto, F. Deep learning-based space debris detection for space situational awareness: A feasibility study applied to the radar processing. *IET Radar Sonar Navig.* **2024**, *18*, 635–648. [CrossRef]
108. Caldas, F.; Soares, C. Machine learning in orbit estimation: A survey. *Acta Astronaut.* **2024**, *220*, 97–107. [CrossRef]

MDPI AG
Grosspeteranlage 5
4052 Basel
Switzerland
Tel.: +41 61 683 77 34

Sensors Editorial Office
E-mail: sensors@mdpi.com
www.mdpi.com/journal/sensors

www.ingramcontent.com/pod-product-compliance
Lightning Source LLC
Chambersburg PA
CBHW061624210926
51850CB00009B/2429
9783725852239